DATE DUE

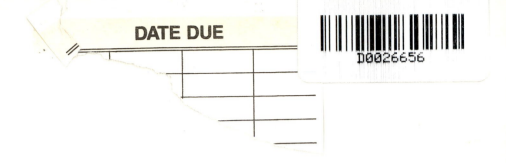

D0026656

The Hydrogeology of Crystalline
Basement Aquifers in Africa

Geological Society Special Publications
Series Editor J. BROOKS

GEOLOGICAL SOCIETY SPECIAL PUBLICATION NO 66

The Hydrogeology of Crystalline Basement Aquifers in Africa

EDITED BY

E. P. WRIGHT
Hydrogeological Consultant
Aston Tirrold, UK
(Formerly hydrogeological adviser to the ODA)

W. G. BURGESS
University College London

1992

Published by

The Geological Society

London

THE GEOLOGICAL SOCIETY

The Society was founded in 1807 as the Geological Society of London and is the oldest geological society in the world. It received its Royal Charter in 1825 for the purpose of 'investigating the mineral structure of the Earth'. The Society is Britain's national learned society for geology with a Fellowship of 7000.

Fellowship is open to those holding a recognized honours degree in geology or a cognate subject and who have at least two years relevant postgraduate experience, or have not less than six years relevant experience in geology or a cognate subject. A Fellow who has not less than five years relevant postgraduate experience in the practice of geology may apply for validation and subject to approval will be able to use the designatory letters C. Geol (Chartered Geologist). Further information about the Society is available from the Membership Manager, Geological Society, Burlington House, London, United Kingdom W1V 0JU.

Published by the Geological Society from:
The Geological Society Publishing House
Unit 7
Brassmill Enterprise Centre
Brassmill Lane
Bath
Avon BA1 3JN
UK
(*Orders:* Tel. 0225 445046)

First published 1992

© The Geological Society 1992. All rights reserved. No reproduction, copy or transmission of this publication may be made without written permission. No paragraph of this publication may be reproduced, copied or transmitted save with the written permission or in accordance with the provisions of the Copyright Act 1956 (as amended) or under the terms of any licence permitting limited copying issued by the Copyright Licensing Agency, 90 Tottenham Court Road, London W1P 9HE, UK. Users registered with Copyright Clearance Center, 27 Congress St., Salem, MA 01970, USA: the item-fee code for this publication is 0305-8719/ 92 $03.50

British Library Cataloguing in Publication Data

A catalogue record for this book is available from the British Library

ISBN 0-903317-77-X

Distributors

USA
  AAPG Bookstore
  PO Box 979
  Tulsa
  Oklahoma 74101–0979
  USA
  (*Orders*: Tel: (918)584–2555)

Australia
  Australian Mineral Foundation
  63 Conyngham St
  Glenside
  South Australia 5065
  Australia
  (*Orders*: Tel: (08)379–0444)

India
  Affiliated East-West Press PVT Ltd
  G-1/16 Ansari Road
  New Delhi 110 002
  India
  (*Orders*: Tel: (11)327–9113)

Printed in the UK by Cromwell Press, Broughton Gifford, Wiltshire

Contents

Preface

This volume has been conceived as a result of recent research programmes funded by the Overseas Development Administration (ODA) of the UK Government and carried out by the British Geological Survey with associated field studies mainly in Malawi and Zimbabwe. Final Reports have been produced for the British Geological Survey Open File Series which also list previous reports by individual team members. The Project Team was made up from staff of the British Geological Survey (BGS) and the Institute of Hydrology, in association with staff of national organizations in the countries concerned.

A Commonwealth Science Council Workshop on crystalline basement aquifers was held in Harare, Zimbabwe in June 1987 attended by participants from the majority of countries in Africa with significant basement outcrop, along with invited speakers from Australia, Brazil, India and Sri Lanka. The Proceedings were published in 1989 (CSC(89)WMR-13, Technical Paper 273). Some results of the BGS project studies were also reviewed at the Geological Society in April 1989, presentations at this meeting forming the basis of this Special Publication. A number of invited papers on regional issues in Africa are also included and are designed to assist an understanding of the feasibility of extrapolation of the results of the more detailed local studies.

The Overseas Development Administration of the UK Government provided the funds for the research programmes carried out by the British Geological Survey and has also been generous in the support of publication costs. The Geological Society meeting was organized jointly by the Society's Hydrogeological Group and by the Association of Geoscientists for International Development (AGID).

From WRIGHT, E. P. & BURGESS, W. G. (eds), 1992, *Hydrogeology of Crystalline Basement Aquifers in Africa*
Geological Society Special Publication No 66, pp 1–27.

The hydrogeology of crystalline basement aquifers in Africa

E. P. Wright

The Baldons, Baker Street, Aston Tirrold, Oxon OX11 9DD, UK

The author was the Project Leader of the British Geological Survey (BGS) Research Projects on Basement Aquifers and Collector Wells. This introductory paper constitutes an overview of crystalline basement aquifer occurrence, geological and climatic controls, evolution, geometry, hydraulic parameters, resources and recharge, combined with a discussion on exploration and management. Data from the BGS Research Projects Report Series are incorporated.

Research programme rationale

Basement aquifers are of particular importance in tropical and sub-tropical regions both because of their widespread extent and accessibility and because there is often no readily available alternative source of water supply, particularly for rural populations. Even in more humid tropical regions, water quality considerations can favour their use.

Basement aquifers are developed within the weathered overburden and fractured bedrock of crystalline rocks of intrusive and/or metamorphic origin which are mainly of Precambrian age. Sedimentary cover rocks, even when consolidated and of comparable age, usually differ in certain key aspects, most notably in mineralogy with a preponderance of components of lower susceptibility to chemical weathering (quartz sands and clays). Unmetamorphosed volcanic rocks can also be distinguished since they may have significant primary porosity and layering along with associated sedimentary intercalations. Additionally, recent volcanic rocks occur mainly in upland areas where any weathering products tend to be rapidly eroded.

Basement aquifers of significant extent in tropical regions occur in sub-Saharan Africa (Fig. 1), South Asia, South America and Australia. Africa is distinctive in this grouping because of the continent's very dispersed rural populations which, in combination with the poor economic base, is responsible for the small scale of existing groundwater development, currently almost wholly for domestic supply or livestock use. Recent estimates are that only some 30% of the rural populations of sub-Saharan Africa have access to a clean potable water supply of minimum quantity (25 litres per head per day). The bulk of the shortfall will have to be met by groundwater, much of which will be derived from basement aquifers.

Because of the typically low productivity of basement aquifers, development is mainly from point sources, utilizing handpumps or bucket and windlass systems. Economic constraints, rather than overall resource availability, tend to limit borehole numbers and it is estimated that development to meet minimum supply standards for rural communities would abstract the equivalent of 1–3 mm/a areal recharge. Present evidence often indicates substantially higher orders of actual recharge, probably in the range of 30–150 mm/a where mean annual rainfall is 500–1000 mm. Although some of this recharge has beneficial effects in relation to phreatophyte vegetation growth and stream baseflow, there is obvious potential for additional direct usage, for improved water supply, both rural and urban, and small-scale irrigation, if methods of increased abstraction can be developed which are both feasible and economic.

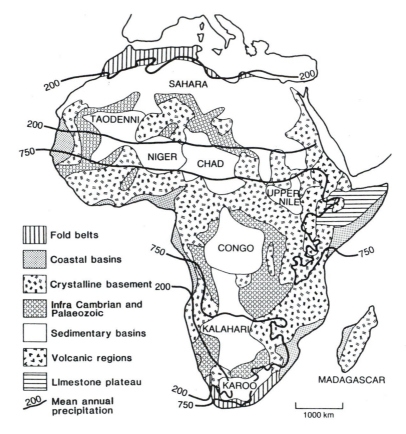

Fig. 1. Groundwater regions in Africa (after Dijon–Les Eaux Souterraines de l'Afrique).

There are a number of important constraints to current development of basement aquifers which include:

(1) the frequent high failure rate of boreholes, commonly in the range 10–40%, with the higher rates in drier regions or where the weathered overburden is thin;
(2) shallow occurrence and fissure permeability of the bedrock aquifer component which makes for susceptibility to surface pollutants;
(3) the low storativity of basement aquifers which may therefore deplete significantly during sustained drought periods. Recharge is also sensitive to certain land use changes, notably those associated with desertification.

The most immediate needs are to improve borehole success rates, reduce maintenance requirements of water supply systems and to develop methodology for increased economic abstraction. However, there is also an important need to evaluate the overall resources and aquifer occurrence more precisely, not only to assist development efficiency but also for longer-term sustained control. Basement aquifers are distinctive in that their occurrence and characteristics are largely a consequence of the interaction of weathering processes related to recharge and groundwater throughflow. A close relationship exists therefore between groundwater occurrence and relief, surface water hydrology, soil and vegetation cover. Improvements in the understanding of these relationships will be fundamental to the planning and management of groundwater resources in crystalline basement terrain and reduction of development costs.

The BGS research programme included some regional assessment reviews and a number of local studies in Malawi, Zimbabwe and Sri Lanka. The objectives of these studies were in general accordance with the research needs listed earlier but mainly concentrated in areas of recent development projects enabling an improved database availability. Malawi and Zimbabwe exhibit important contrasts in basement aquifer occurrence. In Malawi, development is primarily restricted to the aquifer in the weathered overburden. By contrast, in Zimbabwe, boreholes are typically completed in the fractured bedrock, a consequence of the generally thinner overburden.

Basement aquifer occurrence

Basement aquifers occur within the weathered residual overburden (the regolith) and the fractured bedrock. Development of the regolith aquifer component is by wells and shallow boreholes which are able to be drilled by lightweight percussion rigs. Boreholes which are designed to penetrate the fractured bedrock to a significant extent require more powerful drilling rigs, preferably air hammer. Viable aquifers wholly within the fractured bedrock are of rare occurrence because of the typically low storativity of fracture systems ($< 1\%$, Clark 1985). To be effective, development of the bedrock component requires interaction with storage available in overlying or adjacent saturated regolith, or other suitable formations such as alluvium.

Basement aquifers are essentially phreatic in character but may respond to localized abstraction in semi-confined fashion if the rest water level occurs in a low-permeability horizon, such as clayey regolith. Although the aquifers have a regional occurrence, they respond to abstraction in 'discontinuous' fashion, due either to discontinuities or barrier boundaries within the fracture system being tapped or to the constraints of the low-permeability regolith. These features are commonly reflected in a significant borehole failure rate and a wide range of yields, despite the apparent regional uniformity of the basic controls of climate, morphology and geology.

Recent studies in Europe and America in association with radioactive waste disposal have increased knowledge of the detailed hydraulics of bedrock fracture systems (Black 1987). Most other information on basement aquifer occurrence is derived from groundwater development projects. The high failure rate of boreholes testifies to the difficulties of predictive exploration in this environment. The cause must lie in the limited sensitivity of current exploration techniques and an incomplete understanding of the controls on basement aquifer occurrence.

Weathering profile

Several zonal groupings have been described in the literature for the lithological sequence above crystalline basement rocks in tropical regions, variably based on physical, chemical or mineralogical features (UNESCO 1984). The units of the BGS classification (Fig. 2) have a mainly physical basis but with associated mineralogical features. Various subdivisions of the three main zones can often be recognized.

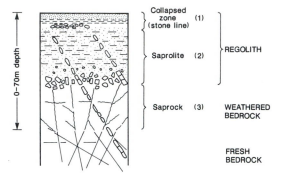

Fig. 2. Schematic weathered profile above crystalline basement rocks. (1) Collapsed zone. This may show marked lateral variation but is generally sandy on watershed areas with illuviated clay near the base and sometimes a 'stone line'; on valley slopes, colluvial material accumulates and in valley bottoms, secondary clay minerals predominate. Slope bottom laterites may also occur which can result in perched water tables. Permeabilities vary in accordance with lithology although on watersheds the collapsed zone normally occurs above the water table. (2) Saprolite is derived by in situ weathering from the bedrock but is disaggregated. Permeability commonly increases at lower levels due to a lesser development of secondary clay minerals. (3) Saprock is weathered bedrock. Fracture permeability generally increased as result of weathering (as compared with fresh bedrock) unless infilled by illuviated clay minerals.

Bedrock. This includes the variably fractured fresh bedrock and the saprock or weathered bedrock. The saprock–fresh bedrock junction is generally transitional or even fluctuating in banded sequences. Fracture systems are related either to decompression or to tectonic forces. The former tend to be subhorizontal with a decreasing frequency with depth. The latter tend to be subvertical and are often in zonal concentrations. Since tectonic fracturing in basement rocks has occurred on several occasions over a long time span and often with reactivation of old fractures, clear recognition of age or differentiation of shear and tensile conditions is not to be expected.

Fissure permeability is assumed to correlate to some degree with frequency of fracture occurrence, with a further assumption that both parameters will decrease with depth. However, the few detailed studies do not always present comparable features, variations being probably due to limitations of the database. Data from boreholes in southern Zimbabwe (Houston & Lewis 1988) showed a greater frequency of fractures in the upper 20 m of bedrock below the regolith interface. Such a distribution probably reflects fracture systems mainly of decompression type. Fracture sealing in the weathered bedrock profile is known to occur, probably by clay illuviation, and may result in an increasing permeability with depth (Megahan & Clayton 1986). Fracture occurrence and permeability in the main bedrock may show

Table 1. *Hydraulic parameters in fissured crystalline rocks*

Country	Rock type	Borehole numbers	Mean transmissivity $(m^2 d^{-1})$	Range transmissivity $(m^2 d^{-1})$	Range apparent permeability $(m d^{-1})$	Comment
Zimbabwe	Mobile Belt Gneiss	228	4.2	0.5–79	0.01–2.3	Pumping test*
	Younger Granite	309	3.6	0.5–71	0.01–1.9	Pumping test
	Older Gneiss	392	—	0.5–101	0.01–2.8	Pumping test
Malawi	Biotite Gneiss	2	—	—	0.1–0.2	Packer tests (1 m)
United States	Granite	58	—	—	0.0–1.7	Injection tests[†]
Europe/USA	Varied	8 sites	—	—	1×10^{-8}–0.9	Straddle packer tests (1–25 m)[‡]

* Estimates based on pumping test data in Zimbabwe Government records. Boreholes typically include thin saturated regolith but with majority deriving principal yields from bedrock.
† Megahan & Clayton (1986).
‡ Black (1987).

an inverse relationship of depth to permeability with a large data set (Carlsson *et al.* 1983) but at a number of intensively studied sites in Europe and the USA, no depth dependency of permeability to 700 m was identified (Black 1987). The depth–permeability relationships at individual borehole sites are obviously affected by the dip of tectonic fracture zones.

Some results of hydraulic testing are shown in Table 1 with apparent permeability values being derived either from packer or pumping tests correlated with penetrated section or packer length. The results suggest that the apparent permeabilities of the more significant fracture systems occurring in production boreholes are in the range 0.01–3 m/day.

Regolith. The regolith consists of the collapsed zone and the saprolite. Since weathering is most effective in the vadose zone and the zone of water table fluctuations, there is a tendency to develop subdivisions into an upper and lower saprolite relative to current (or previous) water levels, in addition sometimes to a basal brecciated zone where rock fragmentation is largely unaccompanied by mineralogical changes.

In regions of moderate to high annual rainfall (>600 mm/a), the water table occurs typically at shallow levels (<10/15 m) and the regolith aquifer provides the main storage for deep boreholes as well as both storage and transmissivity for wells and shallow boreholes.

The collapsed zone has developed from the underlying saprolite by further dissolution and leaching, combined with other formative processes: chemical, physical and biological. It includes the surface soils and other layered features such as laterites, calcretes, illuviated clay layers and stone lines. The surface material is typically sandy on watershed areas where these overlie quartz-rich rocks but changes to sandy clays and clay (montmorillonite) in valley bottom lands (Fig. 3). The surface sands have high infiltration capacities which decrease markedly in any underlying illuviated clay horizons.

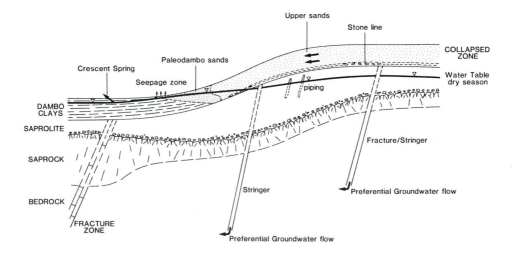

Fig. 3. Schematic basement aquifer catenary cross section.

Saprolite is derived from in situ weathering and is disaggregated. An upper saprolite may be distinguished by higher proportions of the more advanced secondary clay mineral (kaolinite); the lower saprolite has a greater abundance of primary minerals combined with the earlier forms of secondary clay minerals (smectites). The boundary with the underlying saprock may be sharp (against coarser-grained, massive rock) or transitional (against banded or finer-grained rocks).

Regolith thickness and lithology, along with corresponding aquifer hydraulic parameters, depend on a complex combination of controls, including:

bedrock characteristics (chemistry, mineralogy, petrography and structure);
climate (past and present);
age of land surface;
relief and other site factors.

Desipite the common regional coherence of these major controls, the variability of aquifer response to abstraction suggests an over-riding influence of local features which are thought to occur mainly in small-scale lithology or structure or in terrain features affecting leaching history.

The generally greater thickness and the more advanced weathering products in the regolith overlying older erosional surfaces reflect the importance of the duration of weathering. However, the effects of northward continental drift and latitudinal position may also be the cause of significant differences in the regolith thickness, and possibly lithology also, on the same land surface in different parts of the African continent. These differences may account for the apparently higher productivity and greater mean thickness of the regolith aquifers below the African surface of Malawi as compared with Zimbabwe. This is apparent from statistical analysis of borehole data and from normalized baseflow measurements.

Information on the hydraulic conductivity of saturated regolith is mainly available from pumping tests on regolith boreholes and wells and more rarely from packer tests (Table 2). Regolith permeability values in the BGS study areas of Malawi and Zimbabwe are generally less than 0.5 m/day and are significantly lower than the upper ranges in fissured bedrock aquifers. Total porosity values from geophysical logs give high values (20–40%) but data on specific yield are almost non-existent. Values are likely to be low in the more clayey regolith.

The relationship of weathering processes to permeability is complex and variable. Initial weathering which results in the disaggregation and removal of material in solution without the production of secondary minerals must progressively increase permeability and specific yield. Later production of secondary clay minerals will reduce the value of these parameters and clays could seal fractures in the bedrock. If more aggressive weathering should occur at a later stage, this could result in dissociation of kaolinite and increased permeability.

Since weathering is most effective in the vadose zone and the zone of water table fluctuations, the upper saprolite will tend to be more clayey and when saturated will have relatively low permeability and specific yield. The layered permeability is apparent from the semi-confined response of boreholes and can also be deduced from the response of collector wells as compared with the original large diameter well.

Permeability is likely to be low when the regolith is derived from rocks rich in ferromagnesian minerals, notably biotite, which convert easily to secondary clay

Table 2. *Hydraulic data on regolith aquifers in Malawi and Zimbabwe*

Location	Total thickness (m)	Saturated thickness (m)	Transmissivity (m² d⁻¹)	Permeability (m d⁻¹)	Comments
Malawi					
Livulezi	24.1	17.6	1–20	—	134 Bh; mean values*
Dowa West	24.7	16.1	0.2–5	—	81 Bh; mean values*
Lilongwe	30.0	22.2	—	—	18 Bh; mean values*
Mponela North	20	16.6	26	1.5	Collector well†
Mponela South	35	32.2	very low	very low	†
Zimbabwe					
Masvingo area	18	13.3	5.2	—	64 Bh; mean values‡
Masvingo area	29	24	3.3	—	15 Bh; mean values‡
Murape	20	17.4	7.0	0.4	†
Hatcliffe Windpump	26	22.7	12.0	0.5	†
Hatcliffe Willowtree	14	8.1	40.0	4.9	†
St Nicholas	12	7.7	1.4	0.2	†
Mukumba	11	8.1	0.2	0.02	†
St Liobas	17	13.2	1.5	0.11	

* Central Plateau of Malawi, African surface. Data from Chilton *et al.* (1984).
† Collector wells. Data from Wright *et al.* (1988).
‡ Post-African land surface. Data from Houston *et al.* (1987).

minerals. An additional factor is the aggressiveness of weathering. At low weathering rates, biotite converts to expanding hydrobiotite which produces low permeability regolith (e.g. Makumba, Table 2). With more aggressive weathering, such as may occur at the transition zone between successive erosion surfaces (St Nicholas, St Liobas, Table 2), biotite converts directly to vermiculite or kaolinite which results in a relatively higher permeability. A similar correlation has been referred to by Acworth (1987) based on the results of studies by Eswaran & Bin (1978) which demonstrated that low groundwater throughflow rates result in the formation of interstitial clay as compared with higher flow rates which resulted in pseudomorphs. The latter tend to retain the texture and porosity of the original disaggregated regolith and provide higher permeability. Larger hydraulic gradients would therefore be expected to result in higher permeability of the regolith. This is not always apparent from statistical studies but the contrary effects may be masked by increasing thickness weighted by a higher probability of intersecting rapid throughflow channels.

Jones (1985) has extended this concept by suggesting that reduction of recharge could correspond with reduced aquifer permeability by blocking with clay minerals; regolith aquifers in low rainfall areas would, on this assumption, tend to have low permeabilities irrespective of their earlier history. More studies of the mineralogy and permeability of regolith aquifers in arid zone regions are needed to confirm this assertion. However, low saturated thicknesses are also likely to exercise an over-riding constraint to groundwater development in these regions.

The regolith is variably intersected by rapid throughflow channels, some of which are shown in Fig. 3. The overall relative importance of channel to matrix flow is difficult to assess and some evidence suggests that the former invariably predominates; studies by Sharma (1988) in Australia have demonstrated that even in an apparently uniform sand horizon, more than 50% of throughflow is carried by preferential pathways. Rapid throughflow channels include horizontal features such as the stone line at the base of the collapsed zone and the saprolite basal breccia which is an important target for collector wells; and subvertical channels which can include residual quartz bands/veins, biological features such as tree roots and old termite tunnels, and exsolution piping which has enlarged older features such as residual fractures. Residual fractures can occur with a range of trends from vertical to horizontal and are more likely to remain open below the water table. This combination of both horizontal to vertical channelways must enhance both lateral and vertical flow movements. The geochemistry of significant valley spring outflows (high sulphate and low chloride) are often demonstrative of rapid throughflow occurring to considerable depths in the regolith profile.

Regolith and bedrock aquifers are differentiated on the basis of whether the main yields are obtained from one or other of these components, although in both cases, as stated earlier, storage is derived largely from the former. Regolith aquifers occur more extensively where a relatively thick and permeable saturated occurrence exists which is able to meet the demands required, usually of a handpump installation. The central plateau region of Malawi on the African land surface commonly fulfils these conditions. The basement aquifers of Zimbabwe are mostly of bedrock type, particularly below the Post African land surface because the regolith component is generally too thin and/or too poorly permeable to supply even handpump yields.

Statistical evaluations. The variability of basement aquifer responses has resulted in statistical analyses being used with a view to characterizing particular regions and if possible to identify correlations with features which can be observed (e.g. lineaments) or measured (regolith thickness, electrical resistivity, etc). Separate data sets are necessary for features which cannot be quantified, such as bedrock or aquifer (regolith/fissured rock) type and age of erosion surface. The analyses typically include frequency and regression plots of data sets as well as spatial distributions. Although results to date provide useful guidelines for development planning, including economic considerations, the low regression coefficients of the existing correlations limit their value for individual borehole/well siting or design. Neither statistical analysis nor on-site surveys appear able to identify to a high degree of probability these local controls such as throughflow channels, high permeability basal saprolite, productive fracture systems, etc. which appear to exercise over-riding influence on response characteristics and result in marked heterogeneity within small areas.

Frequency analysis. Specific capacity and groundwater individual dissolved ion contents show log normal distributions whereas other features such as regolith thickness, borehole yield, rest water levels and borehole depths have normal distributions, sometimes skewed to the left. Cumulative frequency and probability plots are able to characterize regional occurrence and assist larger scale planning. Yield frequency plots of aquifers from several major bedrock groups in the same region may show comparatively minor differences (Fig. 4) which probably reflect a statistical uniformity of lithology and structure.

Functional relationships

(a) *Structural correlations.* A statistical study of the yields of 163 boreholes in selected study areas of southern Zimbabwe in relation to lineament occurrence is described in more detail by Greenbaum (this volume). The majority of boreholes in Zimbabwe are sited on the basis of a fracture or dyke-related lineament and usually confirmed by geophysical survey. The yield data set examined showed no correlation with proximity to lineaments, with the latter's azimuths or with lineament length. The locational accuracy of most boreholes in the database records is probably inadequate to obtain a reliable assessment of the proximity relationships. An additional factor which could not be taken into account because of data limitations is the possible dip of the fracture zones. The wide range of azimuth trends related to successful boreholes is suggestive of a pervasive influence such as erosional unloading operating on all existing fracture systems.

(b) *Regolith thickness.* Table 3 shows selected matrix correlations for log specific capacity data with aquifer occurrence parameters in the Livulezi area of Malawi (African surface) and the Masvingo area (rock groups 4–7, Fig. 4) of southern Zimbabwe (Post-African surface). The Livulezi boreholes were all drilled to the base of the regolith. The absence of correlation with regolith thickness is suggestive of an over-riding influence of local lithological features of the banded gneiss bedrock. The Zimbabwe boreholes which are usually completed within and draw their main yields from the fractured bedrock also showed no correlation with regolith thickness. There is apparent however, in all sets, a negative correlation with borehole depths and saturated depths which appears reasonable on the assumption that drilling is

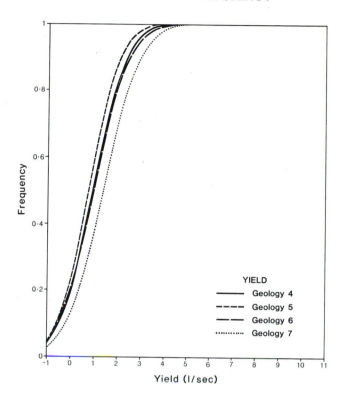

CUMULATIVE FREQUENCY

Fig. 4. Yield frequency plots of aquifers in four major rock groups in the crystalline basement rocks of southern Zimbabwe.

Table 3. *Selected Pearson coefficients*

Log specific capacity	Sample No	Borehole depth	Regolith thickness	Rest water level	Saturated regolith thickness	Depth of saturated aquifer penetrated
Livulezi	135	[−0.23]		0.01	[−0.27]	
Zimb. 4	174	−0.33	0.07	−0.07	0.09	−0.27
Zimb. 5	329	−0.37	0.04	−0.05	0.05	−0.35
Zimb. 6	266	−0.31	0.07	0.08	0.02	−0.39
Zimb. 7	28	−0.07	0.11	0.07	0.12	−0.49

extended either to the base of the regolith (Livulezi), irrespective of thickness or, in Zimbabwe where yields are low, in the hope of striking additional water.

Neither yield nor incremental yield show any correlation with borehole depth or regolith/saturated regolith thickness, as indicated in individual regression plots (Figs 5 & 6). However, an areal plot of the same study area in Zimbabwe (Wright *et al.*

1989), in which all data sets are incorporated, shows a general correlation of high failure rates with low regolith thickness (< 10 m) which is suggestive of the increasing influence of the storage control. Areal plots of regolith thickness also show a positive correlation with rainfall.

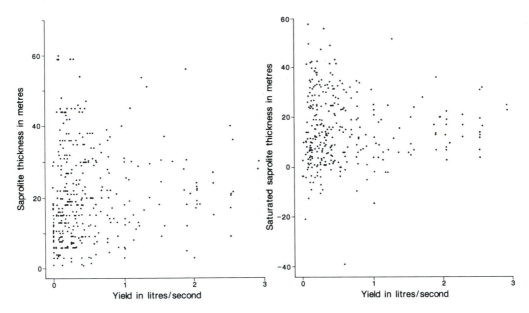

Fig. 5. Regression plots of yields in Porphyritic Granite aquifer of southern Zimbabwe. (a) against total saprolite thickness; (b) against saturated saprolite thickness.

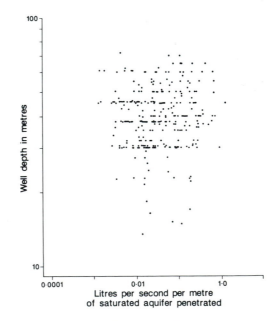

Fig. 6. Incremental yields with depth in Porphyritic Granite aquifer of southern Zimbabwe.

(c) *Terrain correlations.* Terrain correlation studies have been mainly carried out on the central plateau of Malawi which is an African surface, although with eight subdivisions identified by terrain analysis (McFarlane *et al.*, this volume). Yield correlations are mainly apparent in relation to mean values and are negative with relief and positive with regolith thickness. The negative correlation with relief is anomalous since permeability is not apparently reflecting the lower hydraulic gradient; the positive correlation with regolith thickness is contrary to the specific capacity relations in the Livolezi area. The feature may reflect the influence of subvertical throughflow channels, which are increasingly likely to be encountered in thicker regolith sections.

Groundwater resources

Recharge is important not only on account of the small storage of basement aquifers but also because a better understanding of the processes and amounts involved increases knowledge of the aquifer occurrence and potential. This latter point is worth emphasizing since there is a tendency to dismiss the significance of recharge in basement aquifers because the maximum rate of abstraction for rural water supply purposes is likely to be much less than probable recharge.

The basement aquifer, even where continuous, has low permeability and the main groundwater flow systems are relatively localized (Fig. 3) between recharge on watersheds to discharge by runoff or evaporation in valley bottomlands. Basement aquifers predominate below the savanna plainlands of Africa with a mean annual rainfall between 400–1400 mm (Raunet 1985). The so-called bas-fonds (or dambos*) constitute the elementary drainage system of these ferralitized planation surfaces with groundwater flow as the main morphogenetic agent. Although Raunet (1985) demonstrated four main families of bas-fonds in sub-Saharan Africa, there is a basic hydrogeological similarity between them. The area of these bottom lands has been estimated as of the order of 1.3 million km^2. Where mean annual rainfall is less than 400 mm, recharge from direct infiltration is likely to be small or negligible (Edmunds *et al.* 1988) and renewable groundwater resources are only likely to develop in association with runoff either in river alluvium or in underlying regolith sequences.

Rainfall is the dominant control to recharge. The greater part of Africa suffers from rainfall deficit, either chronic and continuous in the desert regions or with seasonal deficits elsewhere and intensified by periodic droughts. Adverse land practices are aggravating desertification and could even have persistent effects on climate. Desertification associated with overgrazing, poor land cultivation practices and deforestation tends to make the land surface more reflective to solar radiation with a consequent reduction in rainfall. Soil crusting by reducing infiltration combined with a reduction in soil moisture storage removal of finer-grained material during soil erosion must also result in reduced evaporation and hence also of such rainfall as is derived from evaporated soil moisture.

The ratio of runoff to rainfall is relatively low in Africa as compared with the majority of other continents (Table 4), a feature which is ascribed to a combination

---

* Dambos are grass-covered, generally tree-less valley bottoms of hydromorphic soils drained by a poorly defined and sometimes non-existent stream channel.

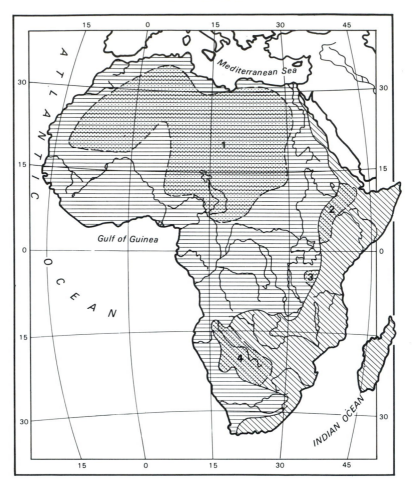

────── boundary of the drainage basins

── ── ── boundary of the areas of internal run-off

drainage basins of the Atlantic Ocean

drainage basins of the Indian Ocean

areas of internal run-off

1   Sahara desert with the basin of Lake Chad

2   basin of Lake Turkana

3   basin of Lake Eyasi

4   Kalahari semi-desert with the basin of the Okavango River

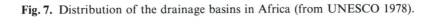

Fig. 7. Distribution of the drainage basins in Africa (from UNESCO 1978).

of lower rainfall, low relief, high evaporation and low groundwater runoff. The latter deficiency is in large part due to low recharge and the low permeability of basement aquifers but also relates to some extent to the localized nature of the aquifer flow systems which promote discharge losses by marginal springs, seepage areas and phreatophyte vegetation.

Table 4. *Ratio of runoff to precipitation in Africa and other main continents**

	Mean annual precipitation (mm)	Runoff coefficient
Africa (Fig. 7)		
Continental mean	725	0.18
Atlantic Ocean slope		
External runoff area	1020	0.24
Internal runoff area	196	0.06
Indian Ocean slope		
External runoff area†	730	0.10
Internal runoff area	648	0.07
Asia		0.40
North America		0.43
South America		0.35

* Data from UNESCO 1978
† Includes Malawi and Zimbabwe

Recharge processes and groundwater flow systems

The combined evidence from lithological, hydrological and hydrochemical data suggests that the basement aquifer in the savanna regions of sub-Saharan Africa consists of a composite flow system with three main interacting components: one near-surface and two with deeper circulation, either through preferential pathways or (slower) matrix flow (Fig. 3). A stream channel in the central bottom land (dambo) is of variable occurrence but becomes more continuous and often incised in a down-stream direction. Groundwater flow resulting from recharge moves laterally towards the dambo and discharges variably in marginal seepage zones, springs and phreato-phytes, and/or via overland flow to a stream channel, or by sub-surface flow downgradient below the dambo clays. Direct losses from the watershed areas can occur by evapotranspiration of deep-rooted vegetation.

The static water levels in high rainfall areas occur typically within the regolith but may fall below the bedrock surface seasonally or in more arid regions. These circumstances have effects on the amount, duration and chemistry of any base flow component of surface flow. Water levels of shallow wells and boreholes completed in the regolith respond in phreatic fashion to recharge although commonly with a marked lag in watershed areas with a substantial thickness of unsaturated zone; in boreholes which penetrate the bedrock, a rapid artesian response may sometimes occur reflecting direct recharge at some distant location.

The near-surface system is associated with high infiltration rates through the surface sands and evidenced by the typical lack of colluvial transfer. Shallow interflow is favoured through high permeability layers such as the upper sands or stone lines, sometimes associated with a perched aquifer. Interflow provides rapid response runoff and also recharge to shallow storage at lower levels such as the palaeodambo sands or the cracks in dambo clays. Vertical infiltration is facilitated by preferential pathways. The geochemistry of the groundwater can assist identification of the nature and depth of circulation which has been evidenced by data from Malawi. Deep but rapid circulation has been identified by low chloride and high sulphate content; shallow interflow by high iron content (McFarlane, this volume).

Measurement of recharge

Methods to estimate recharge are all subject to considerable uncertainty (Simmers 1988) which is generally increased for basement aquifers because of their heterogeneity and the complexity of the flow systems. The studies carried out during the BGS research programme have mainly attempted to improve the understanding of the recharge processes on a regional basis rather than to obtain precise quantification which would have required more detailed local studies. The main methods which have been examined include:

base flow analysis;
estimates of evapotranspirative discharge from dambo seepage zones using climatological data and satellite imagery;
chloride balance of rainfall and groundwater (base flow, springs, boreholes and wells).

Base flow analysis has the advantage of a direct identification of a groundwater component and, where reliable, provides a minimum value of recharge. The hydrological analyses are discussed in more detail elsewhere (Farquharson & Bullock, this volume) but the main features and conclusions may be noted here and some statistics are shown in Table 5. Base flow ranged from zero to 25% of mean annual rainfall or up to 371 mm in absolute terms. There is a dominant rainfall control but analysis and standardization has demonstrated an added correlation with relief and dambo density. Low relief and high dambo density reduce base flow. A corresponding increase in dry season evaporation from the aquifer system is indicated which has been confirmed by process studies using chloride ratios and seepage zone evapotranspirative losses.

Table 5. *Base flow summary data in selected Malawi and Zimbabwe catchments*

Numbers	Mean annual rainfall (mm)	Mean annual runoff (mm)	Mean baseflow (mm)	BF/AR (%)
16 (M)	830–1480	62–487	14–371	0.02–0.25
10 (Z)	551–896	2–155	0–80	0–0.09

With annual rainfall in excess of 800 mm, groundwater recharge appears typically in the ratio range 10–20% which represents a substantial excess over annual demand needs for the current rural populations (1–3 mm).

Dry season evapotranspiration losses which are derived from groundwater can be calculated using the area of the dambo marginal seepage zones and potential evaporation rates (or a proportion). For one catchment in Malawi (5D1) with a 21% dambo cover, calculations indicated dry season losses equivalent to 78 mm over the catchment as compared with base flow measurements of 18 mm. A study aimed at more precise calculations of areal evaporation losses has been carried out on a dambo in Zimbabwe (Stewart 1989) using infra-red data from Landsat TM imagery combined with local ground control. The estimates of actual evaporation were determined at 80% of the potential evaporation rate in the seepage zone and 64% in the clay dambo bottom. The figures can be extrapolated on the TM scene but do also indicate that the previous estimates of seepage zone losses are likely to be of the right order.

The chloride balance method assumes that the chloride content of recharge is determined by the chloride of the effective rainfall concentrated by evaporation and transpiration in the vadose zone. Constraints to the method include uncertainties of the areal rainfall and rapid response runoff but more significantly perhaps in relation to the complex groundwater flow systems in basement aquifer resulting in variations in lag times and the degree of vadose zone concentration. The literature on the method is substantial and several papers in Simmers (1988) discuss case histories or the methodology.

Rainfall chloride contents from Malawi and Zimbabwe from samples taken mainly during the catchment studies are listed in Table 6. The mean chloride content of the Zimbabwe catchments' rainfall is substantially higher than that of samples from Harare and higher than might have been expected from continental interior rainfall. Until confirmed, the results must be regarded with caution.

Table 6. *Chloride content of wet season rainfall (samples of monthly totals) (in mg l^{-1})*

	Number of Samples	$\bar{x}$	Min	Max	σ
Malawi	53	1.01	0.1	5.5	1.2
Zimbabwe	31	2.5	0.9	5.8	1.4
Harare*	8	0.5	0.3	0.8	0.2

* Samples collected 1975–85 by Dr P. Wurzel of the Zimbabwe Ministry of Water Resources and Development.

The chloride contents of surface water runoff in routine monthly samples from several Malawi rivers in the central Plateau region are shown in Fig. 8 (data provided by J. Lewis, Malawi Government Chemist). Wet season values are in the range of 0.6–1.5 mg/l, comparable with rainfall, and dry season values of 6–10 mg/l suggest recharge rates in the range of 10–17% of mean annual rainfall, i.e. 95–162 mm per annum.

Groundwater chloride values typically show a log normal distribution. Recharge is sometimes calculated using the arithmetic mean ($\bar{x}$) although as demonstrated by Eriksson (1976), the harmonic mean ($\bar{H}$) is more appropriate. Some results are shown in Table 7. The harmonic mean tends to give lower values than the arithmetic mean which increases the estimated recharge. Of particular interest are the results of the

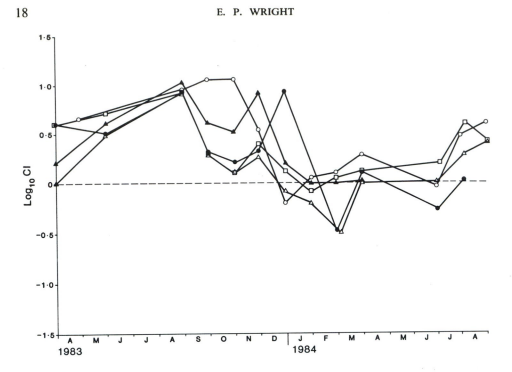

Fig. 8. Chloride content (in mg l^{-1}) of surface runoff in four catchments overlying basement rocks in central plateau region of Malawi.

Table 7. *Mean chloride content of groundwaters (in mg l^{-1})*

Location	Sample	$\bar{x}$	$\bar{H}$
Livulezi	169	7.6	3.7
Dowa Dambo Survey	72	4.5	<1.0
Kampine-Lilongwe Dambo Survey	71	2.1	<1.0
Chimimbe Dambo Piezometers	10	7.5	4.9
Darwendele Tunnel Drips (Zimbabwe)	60	5.2	3.6

two dambo surveys which provide particularly low values of chloride; in the case of the harmonic mean, less than the mean rainfall content. In the Dowa survey, 27% were less than 0.3 mg/l. Samples were analysed at the BGS laboratory to a high standard of accuracy. It may be concluded that the peripheral springs which constituted a major source for sample collection must be derived by transfer through preferential pathways with little concentration occurring. At the same time the dissolved sulphate content of the Dowa samples (93% had sulphate values in excess of 50 mg/l as compared with only 7% in the Lilongwe area) demonstrate the deep circulation of the flow channels leading to the springs. The absence of high concentrations of the other ions demonstrates that the compositions were derived neither by evaporative increase nor solution of near-surface concentrations in a discharge zone.

The difficulties inherent in measurements of recharge rates to basement aquifers make it desirable that several methods should be used for comparative purposes including those which appear most appropriate in the local circumstances. Results based on several methods are shown in Table 8 below. In general the base flow and chloride balance methods in catchments without dambos give comparable results; in those with dambos, base flow estimates are less than either seepage zone calculations or chloride balance estimates but the latter are suspect because of probable rapid throughflow. Chloride data from deep bedrock boreholes show more variable and often higher Cl^- contents than in regolith aquifers and longer duration residence times may be suspected.

Table 8. *Recharge by different methods (mm/a) in Malawi (M) and Zimbabwe (Z)*

Location	Base flow	Cl balance $R(\bar{x})$	Cl balance $R(\bar{H})$	Seepage evapotrans'n	Dambo occurrence	%Mean annual rainfall (%MAR)
Livulezi (M)	145	114	234	—	negligible	10–23
Bua (M)	14	188	—	78–134	high	8–20
Diamphe (M)	75	97	152	150	moderate	12–18
D28 (Z)	80	80	115	–	negligible	9–14

Exploration, development and management

Current development of basement aquifers, being constrained by costs and aquifer productivity, is mainly for rural water supply with substantially lesser usage for urban supply or irrigation. Recharge calculations have indicated rates which are considerably greater than maximum development for rural water supply purposes is likely to exploit. Some of the surplus has potential for other usage taking account of essential demands for base flow, evapotranspiration, or sub-surface irrigation in valley bottom lands. There is scope for management of the shallow aquifers which takes account of both water supply and environmental factors.

Development is by boreholes, dug wells and more recently and to a limited extent, by collector wells. Abstraction may be by motorized or hand/animal pumps or alternative manual devices. The productivity of boreholes or standard wells in basement aquifers often limits abstraction to the two latter methods. An appropriate yield for a handpump is 0.25 l/s and typical success rates based on this criterion in many African countries are around 70–80%. Lower yielding boreholes are some-times equipped with handpumps but seasonal failures are common and maintenance costs tend to be higher. Where such low yielding boreholes are common, extended exploration studies are more justified and a flexible approach to development is recommended to include combinations of boreholes and wells or conjunctive use of groundwater and surface water (rain, runoff).

Regolith aquifers in favourable areas, such as the central plateau of Malawi where there is a thick weathering profile and an annual rainfall of 800–1200 mm, may be developed by boreholes sited by hydrogeological reconnaissance surveys often with a high degree of success. Where success rates or average yields are lower, or higher yields than normal are required, geophysical survey methods may be employed, either vertical electrical soundings (VES) if regolith aquifer development is mainly planned, or remote sensing (air photography/satellite imagery) combined with EM/resistivity traversing if fissure zones are to be targeted. A generally accepted correlation of potential and resistivity for regolith aquifers is shown in Table 9 and graphically of yield and electrical conductance (thickness/resistivity) by Bernardi *et al.* (1988). The correlations are relatively insensitive and inherent sources of error can include over-riding influences of narrow high permeability zones, both subvertical or subhorizontal (basal saprolite) which are not detectable in the overall resistivity response. Statistical data suggest that these influences are not uncommon.

Table 9. *Aquifer potential and resistivity (ohm metres) of layered regolith*

< 20	Clays with limited potential (or saline water)
20–100	Optimum weathering and groundwater potential
100–150	Medium conditions and potential
150–200	Little weathering and poor potential
> 250	Negligible potential

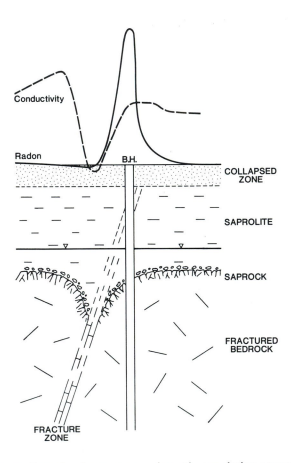

Fig. 9(a). Electromagnetic and radon surveys: schematic correlation across a narrow dipping fracture zone (borehole incorrectly sited).

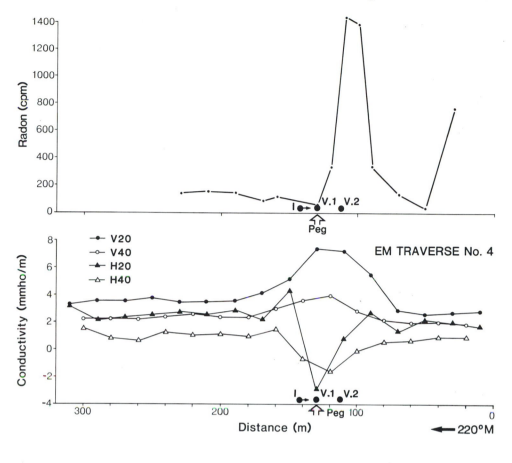

NEMARUNDWE

Fig. 9(b). Profiles across a lineament in southern Zimbabwe demonstrating correlation of electromagnetic and radon responses. Note indication of localized increase in regolith thickness and probable dipping fracture.

The situation is different within the fissured bedrock component. Significant yields may occur from the transitional saprock layer which may be identifiable by VES analysis. Where the saturated regolith is too thin or even in combination with the saprock has too low a transmissivity, deeper fracture zones represent the main target. In these circumstances a flexible approach to development is needed to take account of the limitations of fracture zone occurrence. Wells may be more appropriate than boreholes, such as where thin regolith overlies poorly fractured bedrock. It may be feasible to identify on the land surface, and extrapolate to depth, structural features of likely hydrogeological significance, such as dykes, faults and quartz or pegmatite veins. Their dip may be critical, especially if of narrow width and in areas with deep water levels.

Some exploratory drilling studies were carried out in selected areas with higher than normal failure rates (often > 50%) and in many cases at the general locations of failed boreholes. The results of this drilling demonstrated that significant success and often higher than normal yields could be obtained by more careful surveys, even in locations of thin to negligible regolith cover (Wright 1988). A major emphasis was placed on the precise location of fracture zones by careful examination of air photographs and EM traversing at several spacings. Fracture zones are typically thin in the study areas but it was only in the final analysis of the geophysical data during report preparation that the possible significance of dip was identified, based on the shape of the EM curves. Siting was also assisted by radon gas surveys. These show a close correlation with fracture zone occurrence (Fig. 9) identified by the other surveys. The nature of a radon gas anomaly with its short half-life should serve to confirm the existence of current groundwater throughflow channelling and the scale of the anomaly must bear some relation to the magnitude. The offset between the radon and EM anomaly probably indicates a dipping feature although other factors such as surface clays or recent surface runoff can modify an anomaly. Of the 17 sites drilled, 10 were successful and the 7 failures included 3 sites which would not normally have been drilled since they lacked defined EM and radon anomalies. At least two of the remaining four sites showed dipping fracture zones which could well have affected the actual borehole site drilled. The higher success ratio (71%) is statistically significant at the 10% level in relation to a 50% failure rate in the same areas.

Maximum recommended drilling depths are often based on a subjective assessment of the occurrence of main water inflows as observed during drilling. Optimum drilling depths are determined by economic factors in relation to success rates and yields. Expressed in alternative ways, optimum drilling depths can be either:

(1) the minimum depths at which an adequate yield can be obtained and with a specific capacity which keeps pumps and maintenance costs to a minimum;
(2) the depths at which unsuccessful boreholes should be abandoned so as to give the lowest average cost per successful borehole. Capital and maintenance costs need also to be included.

In regolith aquifers being developed by lightweight drilling rigs, the maximum recommended depth is likely to be the base of the regolith and optimum depths are probably not much less. More uncertainty exists in bedrock aquifers. Some analytical methods have been proposed (Read 1982; Summers 1972), making use of statistically extrapolated surface fracture occurrence with assumed probability distribution at depth. For borehole data from southern Zimbabwe (Wright et al. 1989), evaluations have been based on a statistical comparison of the success rates in different depth intervals and taking account of economic factors. Boreholes are typically completed in the 40–70 m depth range. At 50–70 m depth, minimum success rates of 33–38% are necessary for economic justification (cost break-even). The percentage variation relates to the different bedrock groups. In the event, actual success rates obtained in this interval were between 48–60%.

Collector wells

Collector wells have been studied in the BGS research programme as a means to maximize and optimize abstraction from the surplus resources of basement aquifers. The programme, which included experimental studies and modelling, has demonstrated the feasibility of obtaining substantially larger yields from collector wells than from slim boreholes with the added advantage of low drawdowns. Safe yields from eight wells in Zimbabwe and Malawi and 20 in Sri Lanka have been calculated at 2.7 l/s (ranges of 1.1–6.6 and 0.5–8 l/s respectively) with drawdowns of 2–3 m. These results can be compared with typical yields in the range 0.1–0.7 l/s for slim boreholes at pumping drawdowns in excess of 30 m. The low drawdowns in collector wells make abstraction feasible for low-energy pumping systems—hand, animal, solar and wind. For fuller details of studies including theoretical considerations, construction and testing procedures, and costs, see Wright *et al.* (1987, 1988) and Herbert *et al.* (1988).

The experimental wells have consisted of radial collectors up to 40 m length drilled at the base of large-diameter (2–3 m) wells and completed in the regolith to maximum depths to date of 22 m. The collectors, which can be drilled at any angle required, are mainly targeted on the high-permeability layer within the basal saprolite or upper saprock and analyses of pumping test responses are sometimes indicative of their having interacted with subvertical fracture systems resulting in a higher apparent transmissivity than the theoretical value derived from the large-diameter well data.

A two-dimensional model was able to reproduce the standard Hantush–Papadopulos analysis and a sensitivity analysis was carried out in relation to various critical parameters, such as storage, specific yield, permeability, discharge rates, duration of pumping, well diameter and collector lengths. What appears to be particularly important is the relationship to transmissivity. For high values of the order of 100 m²/day and an aquifer thickness of 7 m, a 2 m diameter well with collectors performs little better than a 3 m well without collectors. For a transmissivity of 5 m²/day, which is fairly typical of the basement regolith aquifer in Zimbabwe, a marked contrast in recovery response is apparent. The differences are demonstrated in Fig. 10 and emphasize the particular applicability of collector wells in low-permeability aquifers.

The model has also been used with data from actual collector well tests. In instances where relatively homogeneous aquifer conditions exist, the model drawdown responses corresponded closely with observed values. In other instances this is not so with the actual well showing a marked 'improvement' on the theoretical, probably as a consequence of the collectors intersecting dipping fracture systems.

Conclusions

The current and future scale of groundwater development for basement aquifers for rural water supply in the context of observed failure rates, variability of yields and maintenance problems, is evidence of the need to increase general understanding of the hydrogeological principles involved. The results of the BGS research programmes have confirmed the range of factors involved in aquifer occurrence and demonstrated the limitations of existing correlations. A recommended planning and development strategy is shown in Fig. 11. Data collection is proposed from a series of monitoring networks which will form the basis for various processing procedures leading to a flexible approach to borehole/well site selection. Although much information can be obtained from routine groundwater development programmes, there is a perceived need for more focussed studies. Better correlations are needed of measurable parameters with aquifer occurrence and in the context of site selection, the recog-

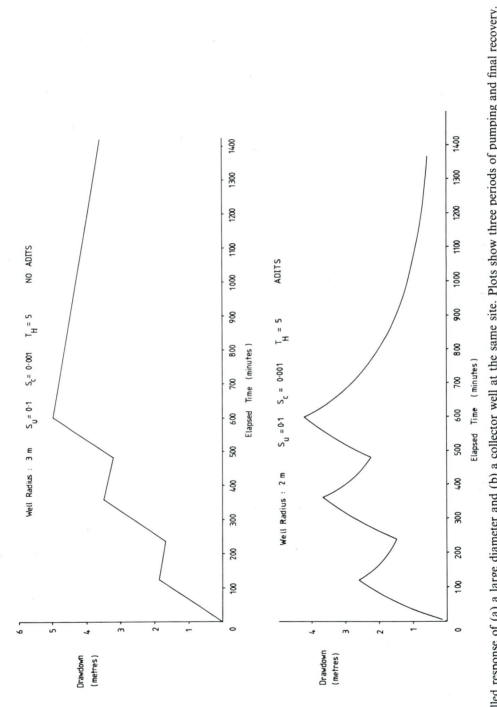

Fig. 10. Modelled response of (a) a large diameter and (b) a collector well at the same site. Plots show three periods of pumping and final recovery.

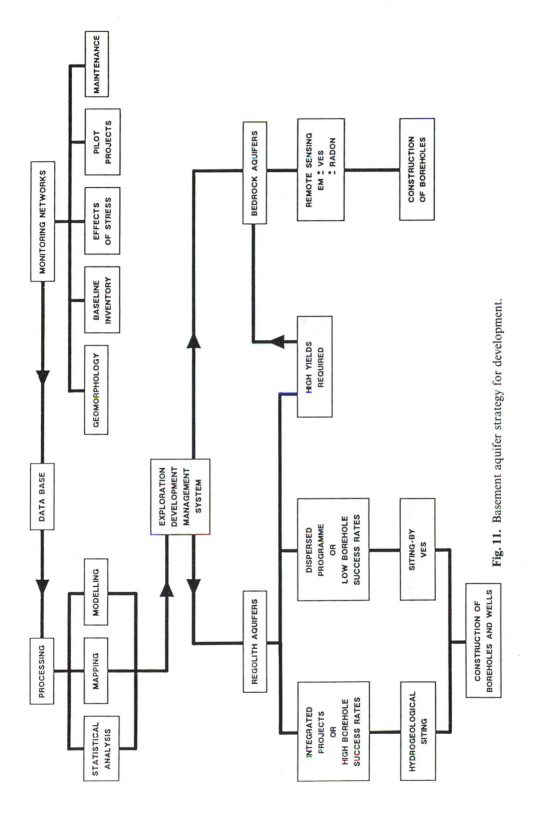

Fig. 11. Basement aquifer strategy for development.

nition and the implications of dipping fracture zones seems worthy of special attention. The use of radon surveys and angled drilling could have a particular application in basement aquifer development. There is also scope for pilot studies on the management of valley bottomlands, including the use of collector wells.

References

ACWORTH, R. I. 1987. The development of crystalline basement aquifers in a tropical environment. *Quarterly Journal of Engineering Geology*, **20**, 265–272.

BERNARDI, A., DETAY, M. & MACHARD DE GRAMONT, H. 1988. Recherche d'eau dans le socle africain. Correlation entre les Paramètres géoélectriques et les caractéristiques hydro-dynamiques des forages en zone de socle. *Hydrogéologie*, **4**, 245–253.

BLACK, J. H. 1987. Flow and flow mechanisms in crystalline rock. *In*: GOFF, J. C. & WILLIAMS, B. P. J. (eds) *Fluid Flow in Sedimentary Basins and Aquifers*. Geological Society, London, Special Publication, **34**, 185–200.

CARLSSON, L., WINBERG, A. & GRUNDFELT, B. 1983. Model calculations of the groundwater flow at Finnsjon Fjallveden, Gidea and Kamelunge. Technical Report 83-45, SKBF-KBS, Stockhölm.

CLARK, L. 1985. Groundwater abstraction from basement complex areas of Africa. *Quarterly Journal of Engineering Geology*, **18**, 25–34.

CHILTON, P. J. & SMITH-CARINGTON, A. K. 1984. Characteristics of the weathered basement aquifer in Malawi in relation to rural water supplies. *In:* *Challenges in African Hydrology and Water Resources* (Proceedings of the Harare Symposium 1984), IAHS Publication 144.

COMMONWEALTH SCIENCE COUNCIL TP273 1989. Groundwater exploration and development in crystalline basement aquifers. Proceedings, Zimbabwe, June 1987, CSC(89)WMR-13.

EDMUNDS, W. M., DARLING, W. G. & KINNIBURGH, D. G. 1988. Solute profile techniques for recharge estimation in semi-arid and arid terrain. *In*: SIMMERS I. (ed.) *Estimation of Natural Groundwater Recharge*, Reidel, Dordrecht, 139–157.

ERIKSSON, E. 1976. The distribution of salinity in groundwaters in the Delhi region and recharge rates of groundwater. *Interpretation of Environmental Isotope and Hydrochemical Data in Groundwater Hydrology*, IAEA Vienna, 171–178.

ESWARAN, H. & WONG CHOW BIN, 1978. A study of deep weathering profile on granite in Peninsular Malaysia. *Journal of the Soil Science Society of America*, 42–49.

HERBERT, R., BALL, D. K., RODRIGO, I. C. P. & WRIGHT, E. P. 1988. The regolith aquifer of hard-rock areas and its exploitation with particular reference to Sri Lanka. *Journal of the Geological Society of Sri Lanka*, **1**, 64–72.

HOUSTON, J. F. T. & LEWIS, R. T. 1988. The Victoria Province Drought Relief Project, II Borehole Yield Relationships. *Groundwater*, **26**, 418–426.

JONES, M. J. 1985. The weathered zone aquifers of the basement complex areas of Africa. *Quarterly Journal of Engineering Geology* **18**, 35–46.

MEGAHAN, W. F. & CLAYTON, J. L. 1986. Saturated hydraulic conductivities of granitic materials in the Idaho Batholith. *Journal of Hydrology*, **84**, 167–180.

RAUNET, M. 1985. Les bas-fonds en Afrique et à Madagascar. *Z. Geomorph. N. F.*, **52**, 25–62.

READ, P. E. 1982. Estimation of optimum drilling depth in fractured rock. AWRC Conference. *Groundwater in Fractured Rock*, Canberra, 191–197.

SHARMA, M. L. 1988. Recharge estimation from the depth-distribution of environmental chloride in the unsaturated zone—Western Australia examples. *In:* SIMMERS, I. (ed.) *Estimation of Natural Groundwater Recharge*. Reidel, Dordrecht, 159–173.

SIMMERS, I. (ed.) 1988. *Estimation of Natural Groundwater Recharge*. Reidel, Dordrecht, NATO Advanced Study Institute, **222**.

STEWART, J. B. 1989. Estimation of areal evaporation from dambos in Zimbabwe using satellite data. *In:* WRIGHT, E. P. (ed.), *The Basement Aquifer Research Project 1984–9*, British Geological Survey Technical Report WD/89/15, 48–59.

SUMMERS, W. K. 1972. Specific capacities of wells in crystalline rocks. *Groundwater*, **10**, 6, 37–47.

UNESCO 1978. *World Water Balance and Water Resources of the Earth.*

UNESCO 1984. *Groundwater in hard rocks.*

WRIGHT, E. P. 1988. *Basement Aquifer Project* (Discussion of radon surveys and exploration drilling in Zimbabwe). Unpublished Report of the British Geological Survey.

——, HERBERT, R., KITCHING, R. & MURRAY, K. H. 1987. Collector wells in crystalline basement aquifers: a review of results of recent research. *In: Proceedings of the XXI Congress of the International Association of Hydrogeologists, Rome, Italy, 12–17 April 1987.*

—— et al. 1988. *Final Report of the Collector Well Project, 1983–88.* British Geological Survey Technical Report WD/88/31.

—— 1989. *The Basement Aquifer Research Project, 1984–89, Final Report.* British Geological Survey Technical Report WD/89/15.

From WRIGHT, E. P. & BURGESS, W. G. (eds), 1992, *Hydrogeology of Crystalline Basement Aquifers in Africa*
Geological Society Special Publication No 66, pp 29–57.

An introduction to the crystalline basement of Africa

R. M. Key

Highlands and Islands Research Group, British Geological Survey, Murchison House, Edinburgh, EH9 3LA, UK

Abstract. The crystalline basement of Africa comprises three major suites of rocks: the granite–gneiss–greenstone association of Archaean cratonic nuclei; strongly deformed metamorphic suites in mobile belts, mainly of Proterozoic age; and anorogenic intrusions which include Phanerozoic intrusive magmatic rocks related to rifting. This heterogeneous basement is extensively concealed beneath a variable thickness of diverse, unmetamorphosed sedimentary and extrusive volcanic rocks and weathering products.

The Archaean cratons have similar geological histories, which generally culminate with anorogenic potassic granites emplaced at about 2500 Ma. The history of the Kaapval Craton is prematurely curtailed at about 3050 Ma. The Limpopo Mobile Belt is uniquely Archaean: its development is linked to differential movement between the oldest cratonic elements of southern Africa. Elsewhere in Africa, the oldest Archaean record suggests an absence of thick continental crust: early greenstone belts formed above mantle plumes on mobile, thin crust.

The development of the Proterozoic mobile belts is related both to collision of older cratons (Wilson cycle orogenesis) and to ensialic disruption of single cratons. Early extensional orogenic phases produced both active (new oceanic crust development) and failed rifts. The earliest Proterozoic mobile belts cut across Archaean cratonic domains and later belts may be superposed. Older fractures are commonly reactivated by new stress systems. Much of the Phanerozoic anorogenic magmatism is related to Mesozoic and Cenozoic continental fragmentation and rifting, possibly associated with hot spots (White & McKenzie 1989).

Notwithstanding the relative antiquity of most of the basement, it is the effects of high-level brittle fracturing and weathering which largely control groundwater storage. The fracturing varies in age from Archaean, within the greenstone belts, to Phanerozoic, for faults related to movement of the African Plate, with ongoing Quaternary faulting in the major tensional rifts such as the East African Rift System. The thickest weathering profiles occur above the oldest erosion surfaces; regional thickness variations are controlled by past and present climatic differences.

The crystalline basement of Africa is composed of metasedimentary, meta-igneous and igneous rocks which vary in age from earliest Archaean to Cenozoic. In the Precambrian crystalline blocks, granite–gneiss–greenstone belts of the Archaean cratonic nuclei are surrounded by essentially Proterozoic orogenic provinces often referred to as mobile belts. Parts of the crystalline basement are igneous intrusions associated with anorogenic magmatism. They include large Proterozoic intrusives such as the Bushveld Igneous Complex, although many were emplaced from Middle Palaeozoic times onwards. Practically the whole of Africa is underlain by Precambrian basement (Fig. 1); on a continental scale the Phanerozoic intrusions are of limited areal extent.

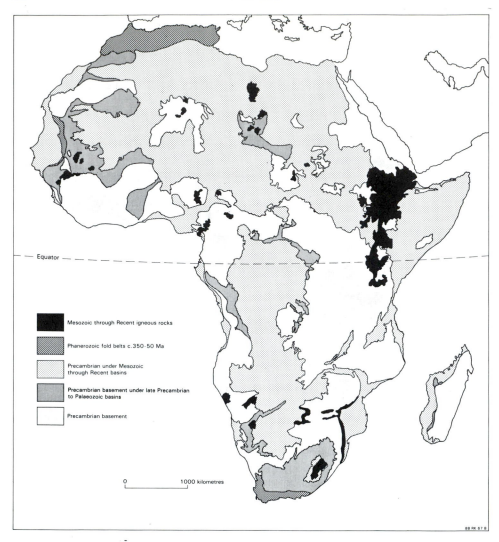

Fig. 1(a). A simplified geological map of Africa showing the distribution of the Precambrian crystalline basement and the main concentrations of Mesozoic-onwards igneous rocks.

The heterogeneous basement is extensively concealed beneath a variable thickness of diverse, essentially unmetamorphosed supracrustal cover rocks. These also vary in age. The oldest cover rocks are the Archaean (about 3000 million years onwards) and Lower Proterozoic sedimentary and volcanic sequences capping the Kaapvaal Craton: the Pongola, Witwatersrand, Ventersdorp, Transvaal–Griqualand West, and Waterberg–Soutpansberg–Matsap Supergroups. The youngest cover sequences include the Cenozoic volcano-sedimentary deposits associated with rifting, notably within the East African Rift System, and the partly consolidated sediments, such as the Kalahari beds, currently infilling the major crustal depressions.

Therefore, in this paper, the crystalline basement refers to intrusive rocks and to recrystallized sedimentary and volcanic rocks within orogenic provinces. At the end of each Proterozoic orogenic cycle, the mobile belts formed part of an enlarged

Fig. 1(b). Political map of Africa.

cratonic area. By the start of the Phanerozoic most of the African crystalline basement had developed.

The crystalline rocks usually possess low primary porosities and permeabilities. Consequently basement aquifers, which include both specific lithologies and fracture zones, tend to be a function of brittle deformation at high crustal levels (see Greenbaum, this volume). Weathering sequences, both ancient and modern, provide some of the most important shallow aquifers. The following lithological characteristics contribute most to the weathered profile to influence aquifer properties: grain size, structure, mineral content and whole-rock chemistry. The most readily weathered basement rocks tend to be coarse grained, badly fractured and rich in high-temperature (ferromagnesian) minerals, with deep weathering commonly concentrated above major fractures. There is a direct relationship between aquifer chemistry and host lithology composition due to water–mineral reactions.

Detailed studies, such as those described later in this book, identify the major basement aquifers on a local scale. The purpose of this introductory paper is to outline the full distribution of the main basement lithologies and structures in order that the local data sets can be extrapolated on a regional scale. In some areas, such as the basement exposures of Malawi, too much local lithological and hydrogeological variation prevents any meaningful regional extrapolation.

It should also be noted that the basement/cover interface, especially if it preserves a weathered palaeosol, is also an important aquifer (exploited at depths to 150 m). The basement acts as an aquiclude with groundwater trapped in the immediately overlying cover rock. In these situations the hydrogeologist is most concerned with the depth to basement and with the nature of the cover material.

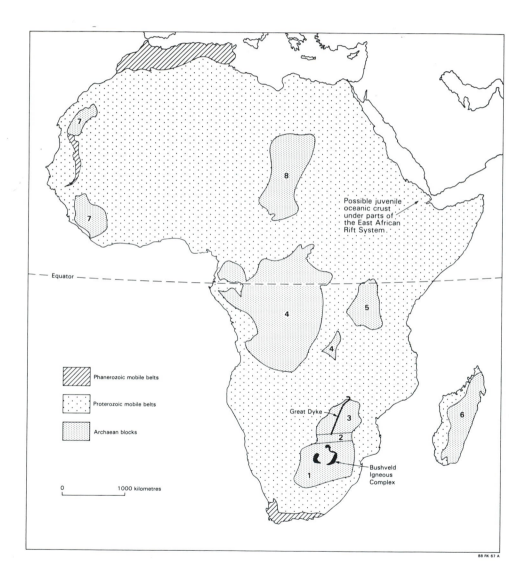

Fig. 2. Archaean blocks which consist of cratonic nuclei and the Limpopo Mobile Belt. 1, Kaapvaal Craton 2, Limpopo Mobile Belt 3, Zimbabwe Craton 4, Kasai or Congo Craton 5, Tanzanian Craton 6, Reworked Malagasy Craton 7, West African Craton 8, North African Craton.

Cahen *et al.* (1984) provide an excellent account of the evolution of the basement of the whole continent based on a complete review of the geochronological record. The geology of West Africa is described by Wright *et al.* (1985). Tankard *et al.* (1982), and sections in Hunter (1981), provide overviews of the southern African basement. References cited in these works are generally omitted in the present paper which has a reference list essentially confined to the most recent work.

Archaean cratonic nuclei

The continental distribution of the Archaean cratonic nuclei is roughly shown in Fig. 2 with the geological histories of the main nuclei summarized in Table 1. Large segments of the Kasai or Congo Craton and of the north African cratonic nuclei (Fig. 3) are covered, mainly by unconsolidated Cenozoic deposits. This means that their geological histories and areal limits are imperfectly known. However, although the western part of the southern African Archaean province is concealed by up to 200 m of Kalahari beds, its main geological components have been established by means of extensive geophysical studies and subsequent confirmatory drilling (*see* Reeves 1978; Meixner & Peart 1984). Detailed geological mapping and geochronological studies have established that all the Archaean cratons have been reworked, at least marginally, during Proterozoic orogenesis. For example, Sanders (1965) showed that the eastern margin of the Tanzanian Craton in Western Kenya was affected, over a width of about 100 km, by thrusting and metamorphism (to produce the Turbo migmatites) related to Upper Proterozoic orogenesis in the Mozambique Orogenic Belt.

The principal components of the Archaean cratonic blocks (excluding the Limpopo Mobile Belt, described separately) are the predominantly low-grade greenstone belts; extensive areas of high-grade gneisses; granite series, including several phases of migmatites, and usually ending with anorogenic K granites; and late minor intrusions (Figs 4 & 5).

Greenstone belts

Two sequences of greenstone belts are generally recognized in the major cratonic nuclei except the Kaapvaal Craton which prematurely stabilized (at about 3050 Ma) prior to the formation of the second generation of belts. The oldest greenstones were laid down between about 3550 Ma and about 3050 Ma; they commonly have precursor gneiss foundations which include definite metasedimentary components. Within these greenstones there are essentially single cycles from basal, mainly basic, volcanics with diagnostic high-MgO rocks (komatiites), upwards into clastic sediment-dominated sequences. They are best preserved on the Kaapvaal Craton and central parts of the Zimbabwe (Rhodesian) Craton. The accounts of Viljoen & Viljoen (1970), Paris (1987) and de Wit *et al.* (1987) of the Barberton Greenstone Belt from the Kaapvaal Craton serve as definitive descriptions of the lithological content of the older belts. Unusually the older (Nyanzian) volcanics of the Tanzanian Craton may have a higher proportion, up to 75% of the volcanic pile, of andesites (Condie 1981), although it is doubtful if the lower part of the Nyanzian is ever seen.

The younger greenstone belts were laid down between about 2800 Ma and about 2600 Ma. They appear to be slightly older in the West African Craton relative to the

Table 1 *A summary of the geological histories of the main Archaean cratons*

(Ma)	Kaapvaal Craton	Zimbabwe or Rhodesian Craton		Tanzanian Craton	Anorogenic granites
2500		Great Dyke emplaced into craton			
2600	Sedimentary and volcanic cover rocks	'Younger' volcano-sedimentary greenstone belts with polyphase tectonothermal history (see text)	Granite series includes migmatites and culminates in anorogenic K-granites	High-grade tectono-thermal event	
2700	Post-tectonic granites emplaced into crustal swells with progressive migration to NW with decreasing time			'Younger' sedimentary volcanic greenstone belts (of more than one generation?) with polyphase tectono-thermal activity	Granite series with no systematic change in K_2O/Na_2O ratio with time
2800					
2900	Granodiorites	High-grade thermal event Greenstone belts	"Granites"	High-grade tectono-thermal event	
3000 3100	----Uplift and erosion of stable craton----			Volcano-sedimentary greenstone belts	
3200	Granite series includes migmatites and culminates in K-granites				
3300	Volcano-sedimentary greenstone belts with polyphase tectonothermal activity producing ENE-trends to schist relicts	central proto-cratonic nucleus stabilized	Granite Gneiss complex: in part high grade meta-sediments	High-grade tectono-thermal event	
3400		'Older' volcano-sedimentary greenstone belts with polyphase tectonothermal activity			
3500	Gneiss complexes: in part high grade metasediments				
3600	>90% granitoids/gneiss <10% greenstone belts	80% granitoids/gneisses 20% greenstone belts		80% granitoid/gneisses 20% greenstone belts (mostly in northern half of the craton)	

Table 1 (cont'd)

Kasai or Congo Craton	N.E. Africa cratonic relicts	West Africa Cratons	(Ma)
K-granites & quartz monzonites			
			2500
High-grade tectono-thermal event	High-grade (amphibolite facies) tectonothermal event (not seen in Tuareg Shield)		2600
Migmatites		Granitoids — High-grade, tectonothermal event to give N-S trends	2700
Tonalites — 'Younger' volcano-sedimentary greenstone belts		'Younger' sedimentary-volcanic greenstone belts	2800
High-grade tectono-thermal event		Granitoid — High-grade tectonothermal event to give E-W trends	3000
Volcano-sedimentary greenstone belts with polyphase tectonothermal activity		'Older' volcano-sedimentary greenstone belts	3100
		Charnockitic basement of uncertain age	3200
High-grade tectono-thermal event (para gneiss basement)			3300
?Tonalite			3400
?Greenstone belt	High-grade gneissification with metasedimentary relicts		3500
			3600
80% granitoid/gneisses 20% greenstone belts	Poorly exposed	80% granitoid/gneisses 20% greenstone belts	

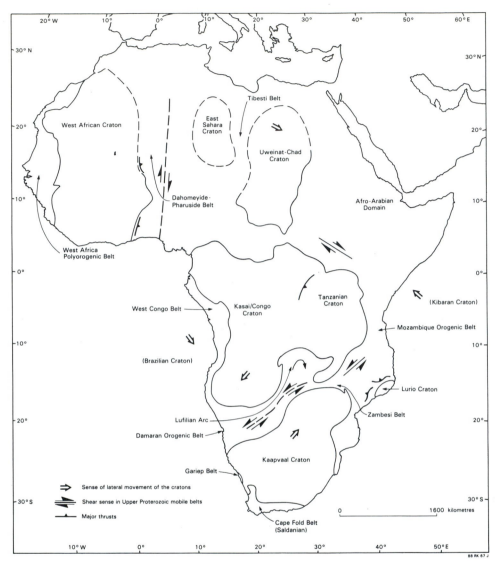

Fig. 3. The inferred full extent of the Archaean cratons is shown together with the relative motion of the Precambrian cratons during Upper Proterozoic orogenesis.

central African cratons although there was a minor development of greenstone belts on the Zimbabwe Craton at about 2950 Ma (Hawkesworth *et al.* 1979). All these belts again comprise single volcanic cycles from basal basic lavas up into more felsic pyroclastics. Both bimodal and calc-alkaline volcanic sequences are recognized. Bimodal assemblages are found in the basal parts of younger belts and contain abundant mafic and ultramafic rocks with minor felsic volcanics and cherts and very little andesitic material. Upper volcanics in younger belts have calc-alkaline affinities and vary from ultramafics through andesites to felsic rocks with associated grey-wackes. Figure 6 illustrates a typical lithological variation within one of the younger greenstone belts of the Zimbabwe Craton (based on Litherland 1975). Mineral variations are used to distinguish up to six types of amphibolite (altered mafic volcanics) by Litherland (1975). However, they have similar whole rock chemistries which closely correspond to oceanic tholeiitic basalts.

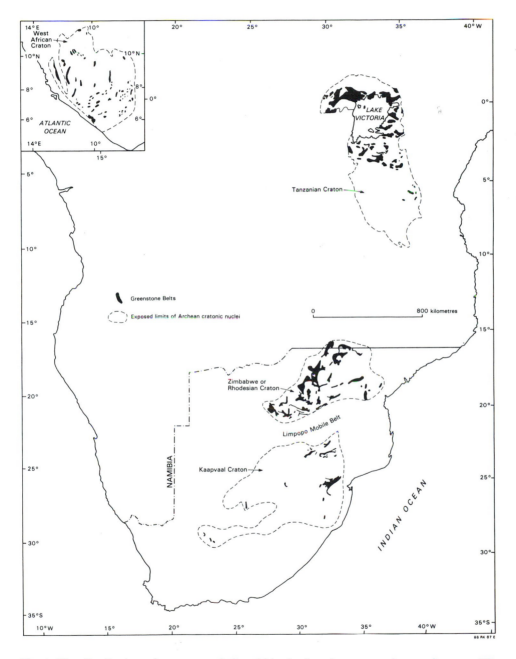

Fig. 4. The distribution of greenstone belts within the four best exposed cratonic areas. The greenstone belts in the northern half of the Tanzanian Craton include the Buganda Supergroup rocks of uncertain age.

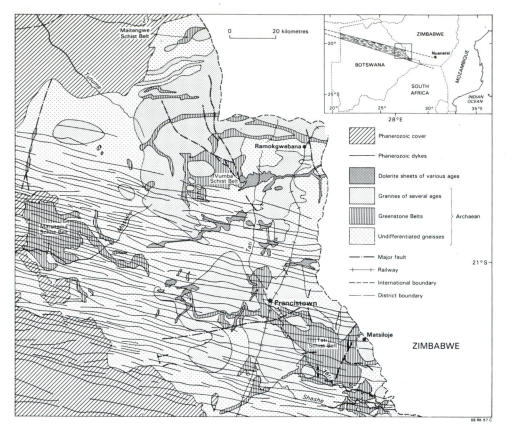

Fig. 5. The geology of a part of the western edge of the Zimbabwe Craton in NE Botswana to illustrate the distribution of the main lithologies and major faults.

Sedimentary sequences are important in the youngest greenstone belts e.g. the Shamvaian group on the Zimbabwe Craton, the Kambui Supergroup of West Africa (MacFarlane *et al.* 1981); the Kavirondian of the Tanzanian Craton; the upper Kibalian of the Congo Craton. The sediments consist of intercalated beds and lenses of chemical and clastic deposits which form highly variable proportions of greenstone belts within individual cratons. Thus, although the average proportion of metasediments within the younger greenstone belts of the Zimbabwe Craton is about 15% (Condie 1981), the Vumba greenstone belt contains minor metasediments (Fig. 6) while the adjacent Tati greenstone belt has major metasedimentary formations (Key, 1976). Typical metasediments in the greenstone belts are Algoma-type banded iron formations, marbles, calc-silicates, metaquartzites, coarse clastic rocks (conglomerates, arkoses etc), aluminous shales, black shales, greywackes and reworked volcaniclastics. These show wide grain-size variations and are chemically varied. The ironstones have along-strike facies variations from chert-hematite/magnetite associations into carbonates and sulphides.

Typical greenstone belt mineralization (i.e. gold disseminated in the metavolcanics or concentrated in fracture-controlled veins; volcanogenic base metal deposits) have primary and secondary effects on groundwater. Base metal concentrations locally

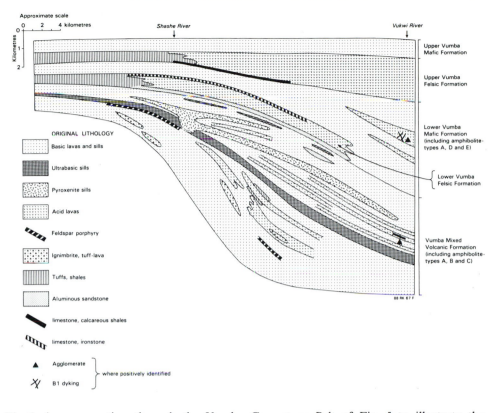

Fig. 6. A cross section through the Vumba Greenstone Belt of Fig. 5 to illustrate the proportions of different volcanic and sedimentary lithologies (after Litherland, 1975). Amphibolite types A and B are lavas; C and D are sills; E is dyke material.

influence groundwater chemistry: oxide-facies gossans derived from sulphide mineralization occur above the water table whose level fluctuates with time. Old gold mine workings in all the greenstone belts are commonly converted to water wells, to provide a secondary useful effect of mineralization.

The greenstone belt terrains have distinctive hilly landscapes controlled by the varied bedrock. Small localized drainage basins are developed with weathered profiles from 15 to 60 m deep (in southern Africa).

Figure 4 shows the distribution of the greenstone belts in the four best-exposed cratons. They are least common in the Kaapvaal Craton, where only the oldest are represented, and most common within the Zimbabwe Craton and northern half of the Tanzanian Craton. The belts are broadly linear throughout the West African Craton and are of higher metamorphic grade (up to granulite-facies; Williams 1978). Within other cratonic domains, the greenstone belts have only suffered greenschist facies metamorphism apart from marginal zones at amphibolite facies. It is possible that the high-grade west African greenstone belts represent disinterred basal remnants. The varied distribution of the greenstone belts apparent in Fig. 4, up to a maximum of 20% by area of the cratons, may be due to a combination of tectonic disruption and variable erosion. At deeper crustal levels granitoid rocks may dominate, especially if the greenstones are compressed within tight synclinal folds.

The granite series (including gneisses) and late minor intrusions

Granites, roughly contemporaneous with spatially associated greenstone belts, are recognized in the main cratonic nuclei. Two main granitic series are recognized, one encompassing igneous activity between about 3600 Ma and about 3100 Ma and the second between about 2950 Ma and 2450 Ma.

The older series commenced with high-grade migmatites which are certainly as old as the adjacent greenstone belts (cf. the Ancient Gneiss Complex of Swaziland; Hunter 1970), or older as the basement in the central African cratons. Metasediments and orthogneisses are present in the early migmatites which are recognized on all'the cratonic nuclei (Table 1). However, the succeeding intrusions have only been mapped and placed into a chronological order in the southern African cratons. Here various major synorogenic tonalitic and trondhjemitic intrusives cut the early migmatites and older greenstone belts and were succeeded by anorogenic potassic granite plutons.

The early sequence is repeated by the second granite series, characterized by calc-alkaline trends, which is much more widely recognized (see Table 1). The migmatites which floor younger greenstones generally record ages of about 2950 Ma (see Cahen *et al.* 1984) or they are slightly younger (Zimbabwe Craton; Baldock & Evans 1988). The succeeding granitoid intrusives generally show progressive increases in K_2O/Na_2O ratios from early tonalitic plutons to anorogenic potassic granites (Litherland 1975; Key *et al.* 1976). The relatively sodic, early rocks underlie featureless plains whereas the later G3 plutons form positive outcrop features, locally with a thick saprock. The relatively high potassium content and the abundance of quartz does mean that this saprock is not broken down into a thick soil cover. Adjacent to fractures the saprock itself may be a good aquifer.

The emplacement of the potassic granites generally marks the end of Archaean orogenesis. This was a diachronous process, from about 3050 Ma (Kaapvaal Craton) to about 2600 Ma for the Zimbabwe Craton and about 2450 Ma for the central African cratons.

Tectonothermal events

Complex vertically plunging structures dominate the early (3600–3200 Ma) African cratonic areas. However, detailed studies of the younger Archaean cratonic areas have revealed polyphase tectonothermal histories similar to those established for Phanerozoic orogenic belts (e.g. Litherland 1973; Stowe 1974). Regional folding produced nappes followed by static metamorphism and emplacement of tonalitic plutons into folded metasedimentary and metavolcanic rocks. After these early events the greenstone belts were isolated as relatively low-grade schist relicts within higher grade gneisses. Further ductile and subsequently brittle tectonothermal events were roughly contemporaneous with the final phases of the granite series. Although the final tectonothermal events were relatively weak compared to higher-grade earlier events they have a critical influence on groundwater storage. These late events generated open folds and crenulations in addition to brittle faults and fractures which are locally important aquifers. Retrogressive metamorphism produced hydrous mineral phases which made the host rock more susceptible to weathering. The average regolith thickness over the Zimbabwe Craton is about 18 m, and is generally from 10 to 30 m in west Africa.

Undoubtedly there were unique features to Archaean geology caused by secular changes to the lithosphere. The older greenstone belts are thought to have originated above mantle plumes, due to the existence of hotter, thinner and more mobile crust within ensialic rifts (Condie 1981). However, the recognition of the similarities of the geological histories of the younger (post-3200 Ma) Archaean cratons and Phanerozoic orogenic belts has generally led to uniformitarian interpretation of the older provinces (*see* Burke & Sengor 1986 and references therein). For example, the youngest greenstone belts are regarded as fragments of oceanic volcanic terrains accreted to continental nuclei during orogenesis. Consequently the development of the younger Archaean cratons is often likened to that of younger orogenic provinces including the Proterozoic mobile belts recognized in Africa. Tankard *et al.* (1982) describe an evolutionary path from mobile belt to craton with gradual lateral growth of African continental crust throughout the Precambrian. However, this is probably an oversimplification as major disruption of the Archaean cratonic blocks took place during the various Proterozoic orogenies and we cannot know how much continental crust was present by the end of the Archaean.

Strike-slip shears and transcurrent faults, over 100 km in length, are characteristic features of modern lithospheric plates. Their existence indicates relative horizontal movement between adjacent competent crustal/lithospheric segments. Therefore the presence of Archaean shears of comparable length can be used as evidence for large, coherent Archaean crustal blocks. In Africa, the oldest of these mega-shears is found in the Limpopo Mobile Belt where they have a maximum age of about 3000 Ma. A logical follow-on of this argument is that the early greenstone belts of the Archaean areas, which are older than the major shear zones, formed in environments devoid of large stable blocks of continental crust. Their generation cannot therefore be related to Wilson-cycle plate tectonic processes, but they may have originated above mantle plumes (cf. Condie 1981).

The Limpopo Mobile Belt

The Limpopo Mobile Belt was first recognized by MacGregor (1953) and trends ENE for about 690 km with a maximum width of about 200 km (Figs 4 & 7). It separates the Kaapval and Zimbabwe Cratons and is dominated by high-grade gneisses and lacks the low-grade greenstone belts, tonalitic plutons and anorogenic potassic granite batholiths normally associated with Archaean provinces. Recent comprehensive reviews of the Limpopo Mobile Belt are provided in Hunter (1981) and in Tankard *et al.* (1982). Orogenic development between about 3200 Ma and about 2500 Ma was dominated by differential (vertical/strike-slip) movement between the Kaapvaal Craton and the ancient central areas of the Zimbabwe Craton. Its evolution as a linear buffer zone conforms to the Sutton & Watson (1974) model for Proterozoic mobile bets. The Great Dyke (emplaced at about 2450 Ma) cuts across the Zimbabwe Craton–Limpopo Mobile Belt boundary to provide a minimum age for the stabilization of the southern Africa Archaean Province.

Three tectonothermal zones are delineated parallel to the mobile belt's length with cratonic components in the two marginal zones (Fig. 7). Three major lithological suites ('basement', metasediments, meta-igneous rocks) are tectonically interleaved, and have been repeatedly recrystallized, in the Central Zone of the Limpopo Mobile Belt. A bimodal gneissic basement of dioritic and granodioritic composition contains

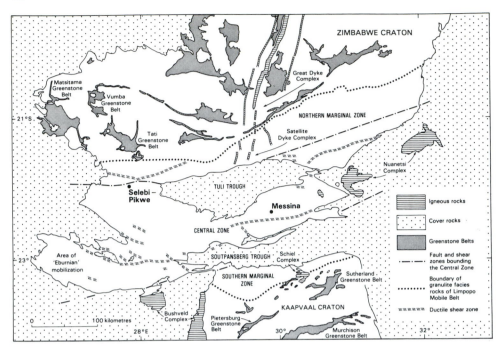

Fig. 7. The main tectonic elements of the Limpopo Mobile Belt.

altered orthogneisses and supracrustal rocks. Several cover sequences are composed of clastic and chemical metasediments as well as metavolcanics (amphibolites) with massive sulphide mineralization. In the Northern Marginal Zone, part of the cover sequence, with granulite facies assemblages, can be traced into the Zimbabwe Craton as part of the younger greenstone belts and their underlying metasediments (Key *et al.* 1976). Syntectonic intrusives are equally varied with major gabbro-anorthosite sheets, granitoids (partly crustal melts) and early dyke swarms. These Archaean rocks are cut by Proterozoic and Phanerozoic dyke swarms and breccia pipes which locally feed overlying lavas within volcano-sedimentary cover sequences. Freshwater springs occur along the interface between the cover and underlying basement at the base of the scarps, notably in eastern Botswana and the northern Transvaal. In these cases the basement is acting as an aquiclude. Aquifers in the basement occur along major faults with long histories of repeated ductile and brittle movement. Polyphase, high-grade tectonothermal events superposed several generations of folding with repeated ductile movement along major shear zones. As such, the Limpopo Mobile Belt can be regarded as a collage of three exotic terranes: the Northern and Southern Marginal Zones and the Central Zone. Subsequent (Proterozoic and Phanerozoic) brittle movement took place along the shears and parallel faults; both sets of structures are important aquifers. Mineral ages of between 2100 Ma and 1770 Ma record the late uplift related to retrogressive metamorphism; the Limpopo Mobile Belt lies within the sourthern part of the area affected by 'Eburnian' orogenesis (next section and Fig. 8). There was contemporaneous crustal remobilization in the western part of the Limpopo Mobile Belt in Botswana (Key 1977).

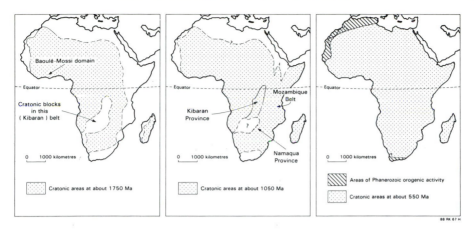

Fig. 8. The known extent of crystalline continental crust, after major orogenesis, at about 1750 Ma, 1050 Ma and 550 Ma. Note that the areas of continental crust shown after 1750 and 1050 Ma were disrupted during later orogenesis.

Crystalline basement development between about 2500 Ma and about 1550 Ma

During this period in excess of two thirds (Fig. 8) of the present African continental crust was affected by a similar sequence of events to those recorded from the Archaean cratonic nuclei. However, controversy remains with regard to the proportion of Archaean material adjacent to the nuclei in the surrounding Lower Proterozoic provinces. This is due mainly to a lack of detailed geological and geochronological knowledge of the Lower Proterozoic provinces, together with poor exposure in many areas, notably northern Africa. However, an increasing amount of isotopic data does imply that a significant amount of new crustal material was introduced around the Archaean cratonic cores (e.g. *see* Taylor *et al.* 1988). A summary of the geological record for this period is given in Table 2.

Low-grade supracrustal sequences are more widely preserved than in the Archaean cratons. The oldest supracrustals are clastic metasediments derived from Archaean cratons during the long period of uplift and weathering at the start of the Proterozoic. They are locally important aquifers and include the altered quartzites, pelites and banded ironstones of the Luiza Supergroup of equatorial Africa and the Oendolongo System of southern Angola, which is dominated by metaconglomerates. The main supracrustal sequences in the 2250–1950 Ma orogenic belts are lithologically similar to those of the Archaean greenstone belts. These include the Birrimian of West Africa (up to 15 000 m thick in Ghana), the Mporokoso Group (up to 5000 m thick) of the Bangweulu Block, and the Buganda Supergroup of equatorial Africa locally interpreted from geochemical evidence, as accreted slabs of ocean crust (Andersen & Unrug 1984). Leube *et al.* (1990) provide an overview of the Birrimian in Ghana. They describe five parallel, evenly spaced, several hundred kilometre-long volcanic belts separated by basins with folded volcaniclastic and clastic sediments as well as granitoids. Although the type area, and the greatest development, of the Birrimian is in Ghana, this supergroup is also exposed in adjacent countries—it covers an area of about 30 000 km^2 in Cote d'Ivoire (Cahen *et al.* 1984). A lower, thick

Table 2. *Summary of the geological record between 2500 Ma and about 1550 Ma (from Cahen et al. 1984)*

Time (Ma)	Geology excludes undeformed platform deposits on Archaean cratonic nuclei	Locality
1500 -		
1600 -	Period of erosion with minor, mostly syenitic intrusives; some supracrustals deposited	Limited data; recorded in Angola and central Africa
1700 -		
1800 -	Stable period: anorogenic alkaline intrusives, dyke swarms, supracrustals locally in rifted environments.	Widespread (see below) although individual intrusions are relatively small
1900 -	Metamorphism	High-grade metamorphic event in central and southern Africa
2000 -	Major orogenesis with several periods of tectonothermal activity with granitoid intrusions. Major sedimentation with volcanic activity. Basal clastic sediments. Major anorogenic intrusives	Areally extensive with fullest geological record in West Africa (Eburnian orogeny with Birrimian sedimentary-volcanic supracrustals) and Angola (major anorthosites). Also recorded in N. Africa (Hoggar, Trans Sahara Belt), Equatorial west and central Africa, central and southern Africa (including Bushveld Igneous Complex). Limited development to include reworking of Archaean cratons, in east Africa
2100 -		
2200 -		
2300 -	Period of erosion with localized sedimentation and tectonothermal activity	Poorly defined tectonothermal event—northern Africa, well defined gneissification—Magondi Belt, central and southern Africa.
2400 -		Mubindji tectonothermal event } Equatorial / Luizian metasediments } West Africa
2500 -	Archaean cratonic nuclei	Emplacement of Great Dyke in southern Africa

sequence dominated by alternating phyllites and greywackes with associated slates, schists and tuffs is overlain by a group of volcanics with minor sedimentary intercalations. Basic lavas and associated intrusives, and less common acidic lavas and pyroclastics, comprise the Upper Birrimian Group. Polyphase deformation was accompanied by greenschist to amphibolite facies metamorphisms. The younger supracrustals are also mixed volcano-sedimentary sequences, commonly deposited in rifts. Thus erosion of the Birrimian volcanics and sediments produced the Tarkwaian Group sediments which were deposited in long narrow intramontane grabens which formed by rifting in the central portions of all five Birrimian volcanic belts (Leube *et al.* 1990).

Associated with the supracrustals is a wide range of intrusions. Alkaline granite series, featuring early large syntectonic plutons are recognized within the main orogens. These include major granodiorites and potassic granites occupying anti-formal zones between synforms defined by Birrimian supracrustals in the Baoule–Mossi Province of West Africa. The large gabbro-anorthosite complexes of southern Angola were also emplaced in the earliest orogenic stages. Migmatites appear to have Proterozoic sedimentary/volcanic rock and Archaean components—most easily recognized in marginal zones to the Archaean cratonic nucleii (cf. Treloar 1988). Post-tectonic igneous activity in the orogenic belts is principally restricted to relatively small intrusions of mixed composition. However, contemporaneous anorogenic magmatism is important within the stable Archaean provinces (see Ng'ambi *et al.* 1986). Both the Great Dyke and Bushveld Igneous Complex were emplaced during the Lower Proterozoic. Dolerite dyke swarms such as the Mashonaland dolerites of central Africa are another distinctive facet of the anorogenic magmatism.

As wide a range of tectonic styles is shown in the Lower Proterozoic ('Eburnian') provinces as in the Archaean cratons. Some have similar sequences of events to the early cratons with initial ductile elements (folds and shears) defining regional structural trends e.g. the NNE grain of the Baoule–Mossi Province defined by the major synforms in the supracrustal relics. Other areas record major strike-slip movement between bounding cratons (e.g. Bassot 1986; Treloar 1988; Andersen & Unrug 1984). The most impressive structures in all the 'Eburnian' provinces are steeply dipping, brittle fractures (in the intrusives) and faults, which may have important hydrogeological potential. The largest faults can be traced for several hundred kilometres, notably in the Tuareg Shield (Fig. 9; Black 1980). These faults may have originated as ductile shears or sutures during the early orogenic history, with repeated subsequent movement to include late brittle faulting, which opened the fractures to groundwater movement. The faults tend to be parallel to the regional trend of the orogenic provinces e.g. N–S to NE–SW within the Baoule–Mossi domain.

The pre-existing stable Archaean provinces must have had a profound influence on the evolution of the 'Eburnian' belts of Africa. It is thought that the lower Proterozoic provinces resulted from either full Wilson-cycle orogenesis, involving collision of separate, relatively small Archaean cratons, or ensialic disruption of a single large craton (Windley 1985; Kroner 1981). However, even the latest detailed studies of single 'Eburnian' provinces (e.g. Treloar 1988) have failed fully to resolve which of these alternative processes took place. This problem is highlighted by Wright *et al.* (1985) in their summary of the main Eburnian province of West Africa. Post-orogenic gravitational collapse and extension of continental crust thickened by tectonic and/or magmatic processes may have produced some mid-Proterozoic sedimentary basins (Andrews-Speed 1989).

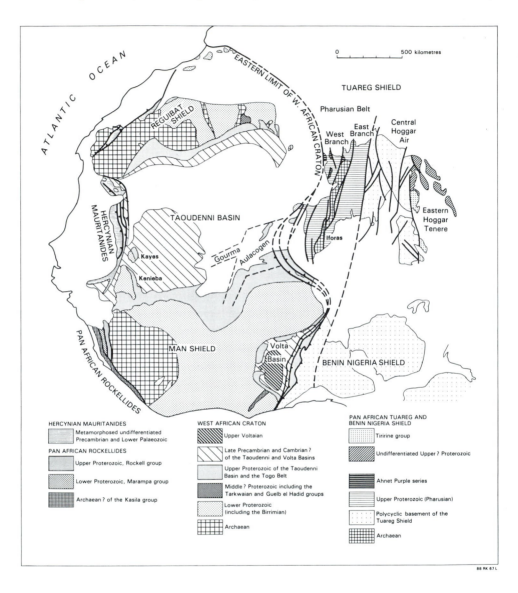

Fig. 9. The distribution of the major faults in the Tuareg Shield (from Black 1980).

Crystalline basement development between about 1550 Ma and 1050 Ma

Orogenic activity was not as widespread as during the preceding period. Two major orogens are recognized; the linear Kibaran Belt of central–west Africa and the arcuate Namaqua Province of southern Africa (Fig. 10; Stowe *et al.* 1984; Borg 1988). The Namaqua Province comprises the Namaqua Belt of South Africa, the

Choma–Kalomo Block and possibly the NE–SW trending Irumide Belt of central–southern Africa. The younger, E–W trending Zambesi Belt separates the Choma–Kalomo Block from the Irumide Belt (Hanson *et al.* 1988). Elsewhere in Africa, less well documented orogenesis took place in the Mozambique Orogenic Belt (El Gaby & Grieling 1988; Key *et al.* 1989; Piper *et al.* 1989). All three provinces are polycyclic with superposed 'Pan-African' events (complete orogenic cycles). A summary of the Middle Proterozoic geological history is shown in Table 3.

A large proportion of the Kibaran Belt comprises metasediments which likely exceed 10 000 metres in total thickness. The supracrustals are dominated by clastic metasediments (important aquifers) with major metaquartzite formations; less common are limestones and greenstones (basic metavolcanics). Metamorphic grade is generally low within this base-metal mineralized belt (Mendelsohn in Hunter 1981). Intrusives include early granitic gneiss complexes as well as composite granitoids such as the Choma–Kalomo Batholith of Zambia (Hanson *et al.* 1988).

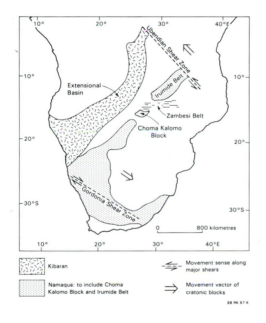

Fig. 10. Major WNW to NW-strike-slip shearing in the Namaqua Province (south) and Ubendian Belt related to extension in the Kibaran Belt.

The Namaqua Province is lithologically more varied with tectonic interleaving of basement gneisses, supracrustals and syntectonic sheet-like intrusions all cut by discordant post-tectonic minor intrusions (Joubert in Hunter 1981; Tankard *et al.* 1982). In this respect it resembles the older Proterozoic crystalline basement provinces. Variable, greenschist to granulite facies, metamorphism associated with tectonic disruption further complicated the lithological diversity. The province is extensively mantled by Upper Proterozoic to Recent deposits. The Irumide Belt in Zambia generally consists of coarse clastic metasediments (Muva Supergroup) with possible felsic metavolcanics. In a western foreland zone these overlie the granitoid Bangweulu Block (Daly 1986). Further east in Malawi a thicker cover sequence is

Table 3. *Geological history between about 1550 Ma and 1050 Ma within orogenic provinces*

Age (Ma)	Event		Location
1150 -	Second period of major orogenesis	Major shearing and strike-slip faulting (central Africa); extensive thrusting e.g. (Mozambique and Natal); syntectonic alkaline granites; post-tectonic mafic dykes and syenites Uplift and cooling, post-tectonic pegmatites and granites throughout "Kibaran" provinces	Central and southern Africa; Egypt, Nigeria and within Mozambique Orogenic Belt
1250 -		Poorly dated sedimentary sequences with or without volcanics e.g. Rio Molocue, Sinclair, Chela, Buanji, Muyumbian, Bukoban Group and Kafue volcanics Post-tectonic granite (Choma–Kaloma Block, Zambia)	Central and southern Africa
1350 -	Main Kibaran orogenesis	Polyphase tectonothermal history with variable metamorphic grade from greenschist to granulite facies. Syntectonic adamellites and Choma–Kaloma Batholith (Zambia) Anorogenic granites	Central and southern Africa, possibly within Mozambique Orogenic Belt of east Africa Angola
1450 -		Stable period over much of Africa; extensional (rift-controlled basins with volcano-sedimentary infill e.g. Kibaran, Burundian, Karagwe-Ankolean Supergroups and Lulua Group of central Africa.	

dominated by metapelites with local carbonates and amphibolite (metagabbro) sheets. These sheets, at least in part, represent altered intrusions and not ophiolite slices. In Malawi, and possibly parts of Zambia, the metasediments are volumetrically subordinate to early granitoid intrusives (Piper *et al.* 1989).

There is geochronological evidence for a Middle Proterozoic basement to the more widespread Upper Proterozoic sediments and volcanics in the Mozambique Orogenic Belt from Mozambique, Malawi, Tanzania and Kenya (Cahen *et al.* 1984; Andreoli 1986; Sacchi *et al.* 1984; Key *et al.* 1989). This basement records a 1100–1200 Ma high grade tectonothermal event. In the north (central Kenya) it is dominated by massive migmatites but a more extensive and varied lithological sequence is described in the south (Mozambique; Sacchi *et al.* 1984). Here, four separate supracrustal sequences have been tectonically interleaved and cut by various granitoid batholiths. The oldest supracrustal formation comprises gneisses and migmatites derived from calc-alkaline volcanics. The younger units are mixed sequences of fine-grained metasediments and metavolcanics which include disrupted ophiolites. The granitoid batholiths which are locally porphyritic, are individually up to about 500 km² in area and form about 25% of the orogenic belt (Fig. 2 of Sacchi *et al.* 1984). Tectonic disruption of the clastic supracrustal components means that they do not form major aquifers.

Two main periods of polycyclic tectonothermal activity have been defined in the main Middle Proterozoic orogenic provinces (Table 3). During both periods the earliest major structures are fold and thrust belts, implying compression across the orogens (*see* Daly 1986*a, b*; Sacchi *et al.* 1984). Ductile shears penetrate through the cover rocks into a crystalline basement which largely controlled the style of deformation. The associated metamorphism locally reached the granulite facies. Subsequent events produced more upright folds and shear zones with large strike-slip movement (e.g. 200 km of dextral displacement across the Gordonia Subprovince in the Namaqua Province; Stowe *et al.* 1984). Contemporaneous strike-slip faulting in adjacent reactivated older belts compensated for shortening in the main orogens e.g. major NW–SE sinistral strike-slip faulting in the Ubendian Belt during oblique compression across the Irumide Belt (Daly 1986).

The recognition of uplifted blocks of basement in the Kibaran Belt influenced early models for the evolution of the Middle Proterozoic mobile belts as ensialic rifts along intracratonic zones of crustal weakness. However, subsequent detailed structural studies in southern and central Africa indicate that the orogenies involved considerable crustal shortening (Matthews 1972; Sacchi *et al.* 1984; Daly 1986*a, b*; Vicat & Vellutini 1988; Gioan & Vicat 1987). Their evolution comprised:

1. Crustal extension. With regard to the early orogenic history of the Kibaran Belt the extension may have produced intracratonic basins without generation of new oceanic crust (Fig. 11; Kroner 1977*a, b*). However, new oceanic crust is recognized in the other provinces (e.g. Andreoli 1984).

2. Crustal shortening to produce fold and thrust belts which tectonically inter-layered sedimentary and volcanic supracrustal rocks and some sialic basement. Major strike-slip movement took place along shears orientated close to the movement vector in the fold and thrust belts. Granulite facies assemblages locally formed (possibly due to crustal thickening), which may have been preceded by granite emplacement in Malawi (Piper *et al.* 1989).

3. Post-collision strike-slip faulting, upright folding and retrogressive metamorphism.

4. Uplift and erosion to commence the next orogenic cycle (of the Upper Proterozoic) in part superposed on all the Middle Proterozoic belts.

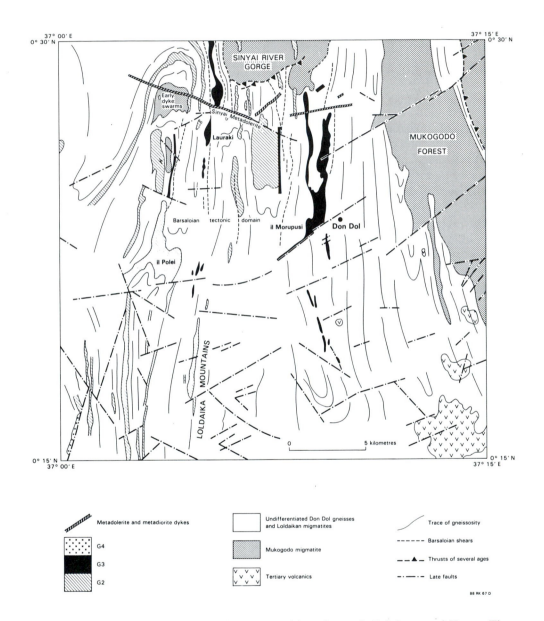

Fig. 11. A geological map of a part of the Mozambique Orogenic Belt in central Kenya. The Mukogodo migmatite is part of an older basement. Note that the late faults do not reflect the N–S grain of the Mozambique belt but mostly trend ENE to NE. These faults are Cenozoic fractures related to stresses associated with the East African Rift System.

Crystalline basement development between about 1050 Ma and about 550 Ma

By the end of this Upper Proterozoic period almost all of the present African continent (Fig. 8c) had formed, and it has remained a stable cratonic area after polyorogenic activity in well defined belts (Fig. 3). Cahen *et al.* (1984) record widespread tectonothermal activity in the orogenic belts at about 950 Ma, 860 Ma, 785 Ma, 720 Ma, 685–660 Ma and from about 600 to 450 Ma. Four major lithological components are variably present in the main orogenic belts, as follows.

Clastic and chemical sedimentary rocks with important fluvio-glacial deposits and stromatolitic limestones (e.g. in the Voltaian and Togo Belt of West Africa, the Limestone and Quartzite Group of Morocco and the Damara metasediments of Namibia). Total thicknesses locally exceed 10 000 metres and important aquifers are present within clastic metasedimentary formations (e.g. the Katangan Supergroup of central–southern Africa). In some cases these rocks are at very low metamorphic grades and should not strictly be regarded as part of the crystalline basement. Their hydrogeological parameters are more akin to the cover aquifers.

Volcanic rocks either as minor intercalations in thick sedimentary sequences or as important volcano-sedimentary provinces tectonically interleaved with the sedimentary sequences. The major volcanic assemblages include the disrupted island arc/ophiolite sequences found in NE Africa (Pallister *et al.* 1988; Shackleton 1986) and NW Africa (e.g. El Graar ophiolite, Leblanc & Lancelot 1980). Alkaline and calc-alkaline volcanic assemblages up to several thousand metres thick are recorded. Again, these volcanic rocks should not strictly be regarded as basement (cf. the sedimentary component).

Intrusive rocks of the alkaline and calc-alkaline granitoid series including syn- and post-orogenic intrusions. The major batholiths are mostly granodioritic (e.g. the early granodiorites of the Mozambique Orogenic Belt in Kenya, Key *et al.* 1989). Major pegmatites are common: e.g. the Khan Pegmatite of Namibia, as are post-tectonic dolerite dykes and sheets (West Africa, Egypt).

Older Basement inliers occur as crystalline foundations at low tectonic levels or tectonically interleaved within the cover sequences (all high-grade orogenic belts). An example of the basement outcrop pattern in the Mozambique Belt of central Kenya is shown in Fig. 11. The basement, Mukogodo migmatite, is exposed in the cores of relatively late antiformal structures. In such complex terrains the main aquifers are the late faults and weathered profiles. There was widespread tectonic reworking of the marginal parts of the cratonic areas (e.g. Treloar 1988; Andersen & Unrug 1984).

Metamorphic grade is variable within single Upper Proterozoic orogens (to blur the distinction between cover and basement with regard to hydrogeology). For example, a range from greenschist to granulite facies assemblages occurs in the Mozambique Orogenic belt of equatorial east Africa and in the Tibesti Belt of north Africa (El Makhrouf 1988). The Upper Precambrian sequences of NE Africa are generally at low metamorphic grades, whereas contemporaneous rocks further south in the Mozambique Orogenic Belt are in the amphibolite or granulite facies. Both terrains are related to the same oblique continent–continent collision. Eroded root zones of the orogen are presently exposed in the Mozambique Belt. Lower grade, higher level parts preserving major slivers of oceanic crust crop out in NE Africa, indicating a lateral change in tectonic style along the orogen. Major strike-slip faulting took place in the north-east (Burke & Sengor 1986). Consequently it is futile to generalize with regard to the lithological make-up of the Upper Proterozoic orogenic belts of Africa (see case studies in later chapters).

The cover sequences of the orogenic belts can be traced onto the cratonic forelands where they are not metamorphosed and are not part of the crystalline basement (e.g. the Voltaian and the Rokel River Group of West Africa). Contemporaneous anorogenic magmatism (e.g. within the cratonic foreland to the Pharusian Belt of north Africa) and major ductile or brittle shearing, such as the Chuan shear zones of the Tanzanian Craton including the Aswa shear zone in Uganda (Hepworth & MacDonald 1966), are also recorded within the cratonic areas between the upper Proterozoic orogens. On the cratons, the intrusions are only of local importance but the shear zones can be traced for up to several hundred kilometres.

All recent authors interpret the development of the Upper Proterozoic orogenic belts in terms of Wilson-cycle plate tectonic processes. Four stages are identified, which may be repeated within a single orogen, as follows.

1. *Rifting*. Initial extension of continental crust (older cratonic areas) with either complete disruption to generate oceanic crust or intraplate, locally transtensional aulacogens (failed rift arms). Some aulacogens are formed by reactivation of old crustal fractures by the new stress fields e.g. the Katangan. The sedimentary infill of these basins may remain largely unmetamorphosed in the aulacogens and form important aquifers (e.g. the Katangan Supergroup). Remnants of the newly formed oceanic crust are recognized, both in low- and high-grade terrains, over the whole of Africa. Porada (1989) shows that the Upper Proterozoic orogenic belts of equatorial and southern Africa evolved from an elaborate system of continental rifts which initially formed on the 'West Gondwanaland' continent about 100–1000 Ma BP.

2. *Subduction and initial collision*. Initial basin closure with accretion of (successive) volcano-sedimentary assemblages onto the cratonic forelands: well documented in NE Africa (see previous references). Major tectonothermal activity giving rise to thrust and fold belts (Coward & Daly 1984) and accompanying magmatism.

3. *Collision between the cratonic fragments (see Fig. 3)*. Continuing tectonothermal activity and magmatism extend into the cratonic forelands. Major strike-slip shear zones within the orogens are aligned subparallel to the trends of the orogens e.g. in the Dahomeyide-Pharuside or Trans Sahara Belt (Champenois *et al.* 1987).

4. *Post-collision cooling and uplift*. Recorded by mineral ages within the orogens and the cratonic blocks. During this period there was a change from subduction-related to within-plate magmatism.

The main aquifers include the thick sequences of clastic and chemical sediments in the aulacogens, weathered profiles, and the major shear zones, both within and outside the polyorogenic belts.

Phanerozoic orogenic activity

The final throes of the Upper Proterozoic orogenesis persisted into the lower Palaeozoic until about 425 Ma. Anorogenic granitoid magmatism and uplift were widespread in the mobile belts. Low-grade tectonothermal activity was mostly localized, although epithermal alteration was ubiquitous in the orogens, facilitating subsequent weathering processes. Renewed major orogenesis in well defined fold-thrust belts was confined to the southernmost tip and northwestern coastal zone of

Africa (Fig. 8c). Elsewhere in Africa essentially unmetamorphosed major Phanerozoic sedimentary basins, e.g. the Karoo of southern Africa, form important components of the cover.

The start of Palaeozoic orogenesis in NW Africa is marked by extrusion of alkaline volcanics at about 560 Ma. Cover sequences of clastic and chemical sediments with volcanics, cut by relatively minor granitoid intrusives, were deformed during Caledonian–Hercynian tectonothermal activity. This produced mountainous fold-thrust belts cut by major wrench faults. The orogeny is a product of the interaction between the North Atlantic and African Plates (Michard & Pique 1979). Major extensional faults and rift-related magmatism during the Triassic and Jurassic preceded Alpine orogenesis in the extreme north.

Inliers of older metasediments and a granite basement are locally present along the mountainous coastal strip, the Cape fold belt which strikes roughly east to west across the southern tip of the continent. A predominantly sedimentary cover of the Middle Palaeozoic Cape Supergroup and overlying mixed Karoo Supergroup dominates this late Hercynian fold-thrust belt which heralded the break-up of Pangaea.

Phanerozoic anorogenic magmatism

The reviews of this topic in Cahen *et al.* (1984), and more recently by Bowden & Kinnaird (1987) and Kinnaird & Bowden (1987), are authoritative. The magmatism, which is widespread, is linked to major faulting and rifting associated with the breakup of Gondwanaland and subsequently (since the Mesozoic) with the development of the East African Rift System. Two major suites of intrusives are recognized as well as the new oceanic crust generated in the Afar area of the East African Rift System. These are the alkaline granitoid ring complexes and basic dykes and sheets.

The alkaline ring complexes are generally sited within the Upper Precambrian mobile belts; emplacement is related to uplift during reactivation of the major shears and transcurrent faults during fragmentation of Gondwanaland. Individual complexes are up to about 100 km in length but mostly from 1 to 30 km in diameter (Batchelor 1987). They tend to form positive topographic features despite extensive epithermal alterations and consequently have limited groundwater potential.

The basic, commonly tholeiitic, dykes and sills are concentrated in areas that remained as stable cratons during the Upper Precambrian (cf. Wright *et al.* 1985) in contrast to the granitoid intrusions. The dyke swarms preferentially weather to control present drainage networks and consequently have important groundwater implications. Water boreholes are commonly successfully sited at the margins of the thicker (up to 50 m) dykes. These dykes are areally important in the main swarms e.g. forming 5% of the area within the late Karoo swarm trending ESE across northeastern Botswana illustrated in Fig. 5. Kimberlite pipes are another small-scale manifestation of the anorogenic magmatism. Reeves (1978) relates the swarm which emanates from Nuanetsi (dated at about 180 Ma), to a failed Gondwana spreading axis.

Post-cratonization effects relevant to groundwater storage in the basement

It is apparent from the other articles in this book specific to groundwater that brittle Precambrian tectonics, Phanerozoic anorogenic events and most importantly, past and present weathering processes, regionally control the groundwater storage capacity of Africa's crystalline basement. Such events include renewed movement along faults and fractures on all scales related to new stress systems created by plate movement. Both major faults within the Upper Precambrian mobile belts and smaller intracratonic fracture zones were reactivated (Guiraud *et al.* 1987; Katz 1987; Pique *et al.* 1987). The anorogenic magmatism influenced groundwater mainly by controlling topography (last section). The interface between crystalline basement and a volcanic carapace is also locally an important aquifer. Groundwater collects at the interface where the basement acts as an aquiclude; commonly the water is discharged at springs at the base of scarps defined by the cover. However, the main influence on groundwater was, and continues to be, weathering related to changing climates. The greatest regolith thicknesses occur above the oldest erosional surfaces. In part these changes were related to the wandering of the African Plate across different latitudes (Van Houten & Hargraves 1987). Variations in the thickness and composition of the regolith across the continent relate not only to past and present climates but also to different bedrock. Consequently it is important for a better understanding of the hydrogeological potential of any part of Africa that the basement geology is properly documented. The fundamental importance to hydrogeological investigations of good geological maps of the basement on as large a scale as possible cannot be over-emphasized.

This paper is based on experience gained while undertaking ODA-funded technical co-operation work in Africa and is published with the permission of the Director of the British Geological Survey (NERC).

References

ANDERSEN, L. S. & UNRUG, R. 1984. Geodynamic evolution of the Bangweulu Block, Northern Zambia, *Precambrian Research*, **25**, 187–212.

ANDREOLI, M. A. G. 1986. Petrochemical, tectonic evolution and metasomatic mineralisations of Mozambique Belt granulites from S. Malawi and Tete (Mozambique). *Precambrian Research*, **25**, 161–186.

ANDREWS-SPEED, C. P. 1989. The mid-Proterozoic Mporokoso Basin, Northern Zambia: sequence stratigraphy, tectonic setting and potential for gold and uranium mineralisation. *Precambrian Research*, **44**, 1–17.

BALDOCK, J. W. & EVANS, J. A. 1988. Constraints on the age of the Bulawayan Group metavolcanic sequence, Harare Greenstone Belt, Zimbabwe. *Journal of African Earth Sciences*, **7**, 795–804.

BASSOT, J. P. 1986. Le complexe volcano-plutonique calco-alcalin de la rivière Dalema (est Senegal): discussion de la signification géodynamique dans le cadre de l'orogenie eburnéenne (Proterozoique inferieur). *Journal of African Earth Sciences*, **6**, 505–519.

BATCHELOR, R. A. 1987. Geochemical characteristics of the Nigerian anorogenic province. *Geological Journal*, **22**, 389–402.

BLACK, R. 1980. Precambrian of West Africa. *Episodes*, **4**, 3–8.

BORG, G. 1988. The Koras-Sinclair-Ghanzi rift in southern Africa. Volcanism, sedimentation, age relationships and geophysical signature of a late Middle Proterozoic rift system. *Precambrian Research*, **38**, 75–90.

BOWDEN, P. & KINNAIRD, J. 1987. Phanerozoic anorogenic magmatism: plate tectonic implications and mineralization. *Geological Journal*, **22**, 293–296.

BURKE, K. & SENGOR, A. M. C. 1986. Tectonic escape in the evolution of the continental crust. Geodynamic Series, American Geophysical Union, **14**, 41–53.

CAHEN, L., SNELLING, N. J., DELHAL, J. & VAIL, J. R. 1984. *The Geochronology and Evolution of Africa*. Clarendon, Oxford.

CHAMPENOIS, M., BOULLIER, A. M., SAUTTER, V., WRIGHT, L. I. & BARBEY, P. 1987. Tectonometamorphic evolution of gneissic Kidal assemblage related to the Pan African thrust tectonics (Adrar des Iforas, Mali). *Journal of African Earth Sciences*, **6**, 19–27.

CONDIE, K. C. 1981. *Archean Greenstone Belts*. Elsevier, New York.

COWARD, M. P. & DALY, M. C. 1984. Crustal lineaments and shear zones in Africa: their relationship to plate movement. *Precambrian Research*, **24**, 27–45.

DALY, M. C. 1986a. The intracratonic Irumide Belt of Zambia and its bearing on collision orogeny during the Proterozoic of Africa. *In*: COWARD, M. P. & RIES, A. C. (eds) *Collision Tectonics*. Geological Society, London, Special Publication, **19**, 321–328.

—— 1986b. Crustal shear zones and thrust belts: their geometry and continuity in Central Africa. *Philosophical Transactions, Royal Society, London*, **A317**, 111–128.

DE WIT, M. J., HART, R. A. & HART, R. J. 1987. The Jamestown Ophiolite Complex, Barberton mountain belt: a section through 3.5Ga oceanic crust. *Journal of African Earth Sciences*, **6**, 681–730.

EL GABY, S. & GREILING, R. (eds) 1988. *The Pan-African belt of northeast Africa and adjacent areas: tectonic evolution and economic aspects of a late Proterozoic orogen*. F. Vieweg & Sohn, Braunschweig/Wiesbaden.

EL MAHKROUF, A. A. 1988. Tectonic interpretation of Jabal Eghei area and its regional application to Tibesti orogenic belt, south Central Libya (S.P.L.A.J.). *Journal of African Earth Sciences*, **7**, 945–967.

GIOAN, P. & VICAT, J. P. 1987. Bilan geochronologique de la Republique Populaire du Congo. *Journal of African Earth Sciences*, **6**, 215–220.

GUIRAUD, R., BELLION, Y., BENKHELIL, J. & MOREAU, C. 1987. Post-Hercynian tectonics in northern and western Africa. *Geological Journal*, **22**, 433–466.

HANSON, R. E., WILSON, T. J., BRUECKNER, H. K., ONSTOTT, T. C., WARDLAW, M. S., JOHNS, C. C. and HARDCASTLE, K. C. 1988. Reconnaissance geochronology, tectono-thermal evolution, and regional significance of the Middle Proterozoic Choma–Kalomo Block, southern Zambia. *Precambrian Research*, **42**, 39–61.

HAWKESWORTH, C. J., GLEDHILL, A. R., WILSON, J. F. & ORPEN, J. L. 1979. A 2.9-b.y. event in the Rhodesian Archaean. *Earth & Planetary Science Letters*, **43**, 285–297.

HEPWORTH, J. V. & MACDONALD, R. 1966. Orogenic Belts of the Northern Uganda Basement. *Nature*, **210**, 726–727.

HUNTER, D. R. 1970. The Ancient Gneiss Complex in Swaziland. *Transactions Geological Society of South Africa*, **73**, 107–150.

—— 1981. *Precambrian of the Southern Hemisphere*. Elsevier, Amsterdam.

KATZ, M. B. 1987. East African rift and northeast lineaments: continental spreading-transform system. *Journal of African Earth Sciences*, **6**, 103–107.

KEY, R. M. 1976. *The Geology of the Country around Francistown and Phikwe, NE and Central Districts, Botswana*. District Memoir Geological Survey, Botswana, 3.

—— 1977. The geological history of the Limpopo Mobile Belt based on field mapping of the Botswana Geological Survey. *Bulletin Geological Survey Botswana*, **12**, 41–60.

——, CHARSLEY, T. J., HACKMAN, B. D., WILKINSON, A. F. & RUNDLE, C. C. 1989. Superimposed Upper Proterozoic collision controlled orogenies in the Mozambique Orogenic Belt of Kenya. *Precambrian Research*, **44**, 197–225.

——, LITHERLAND, M. & HEPWORTH, J. V. 1976. The evolution of the Archaean crust of north-eastern Botswana. *Precambrian Research*, **3**, 375–413.

KINNAIRD, J. & BOWDEN, P. 1987. African anorogenic alkaline magmatism and mineralization—a discussion with reference to the Niger-Nigerian Province. *Geological Journal*, **22**, 297–340.

KRONER, A. 1977a. The Precambrian geotectonic evolution of Africa: plate accretion versus plate destruction. *Precambrian Research*, **4**, 163–213.

—— 1977b. Precambrian mobile belts of southern and eastern Africa—ancient sutures or sites of ensialic mobility? A case for crustal evolution towards plate tectonics. *Tectonophysics*, **40**, 107–136.

—— 1981. Precambrian crustal evolution and continental drift. *Geologische Rundschau*, **70**, 412–428.

LEBLANC, M. & LACELOT, J. R. 1980. Le domaine panafricain de l'Anti-Atlas (Maroc). *Canadian Journal of Earth Sciences*, **17**, 142–155.

LEUBE, A., HIRDES, W., MAUER, R. & KESSE, G. O. 1990. The early Proterozoic Birrimian Supergroup of Ghana and some aspects of its associated gold mineralization. *Precambrian Research*, **46**, 139–165.

LITHERLAND, M. 1973. Uniformitarian approach to Archaean 'schist relics'. *Nature*, **242**, 125.

—— 1975. *The geology of the area around Maitengwe, Sebina and Tshesebe, NE and Central Districts, Botswana*. District Memoir Geological Survey Botswana, 2.

MACFARLANE, A. A., CROWE, M. J., ARTHURS, J. W., WILKINSON, A. & AUCOTT, J. W., 1981. *The Geology and Mineral Resources of Northern Sierra Leone*. Overseas Memoir, Institute of Geological Sciences, UK, **7**.

MACGREGOR, A. M. 1953. Precambrian formations of tropical southern Africa. 19th International Geological Congress, Algiers, 1952, 1, 39–50.

MATTHEWS, P. E. 1972. Possible Precambrian obduction and plate tectonics in south-eastern Africa. *Nature*, **270**, 37–39.

MICHARD, A. & PIQUE, A. 1979. Interpretation geodynamique des Hercynides marocaines en relation avec les mouvements des plaques nord-atlantiques au Primarie. 10 Colloq. Geol. des plaques nord-atlantiques au Primarie. *10 Colloquim African Geology*, Montpellier. Abstracts, 119–120.

MEIXNER, H. M. & PEART, R. J. 1984. The Kalahari Drilling Project: A report on the geophysical and geological results of follow-up drilling to the aeromagnetic survey of Botswana. *Bulletin Geological Survey Botswana*, **27**.

NGAMBI, O., BOERLRIJK, N. A. I. M., PRIEM, H. N. A. & DALY, M. C. 1986. Geochronology of the Mkushi Gneiss Complex, Central Zambia. *Precambrian Research*, 279–295.

PARIS, I. A. 1987. The 3.5 Ga Barberton Greenstone Succession, South Africa: implications for modelling the evolution of the Archaean Crust. *Geological Journal*, **22**, 5–24.

PALLISTER, J. S., STACEY, J. S., FISCHER, L. B. & PREMO, W. R. 1988. Precambrian ophiolites of Arabia: geological settings, U-Pb geochronology, Pb-isotope characteristics and implications for continental accretion. *Precambrian Research*, **38**, 1–54.

PIPER, D. P., CHIKUSA, C. M., CHISALE, R. T. K., KAPHWIYO, C. E., MALUNGA, G. W. P., NKHOMA, J. E. S. & KLERKX, J. M. 1989. A stratigraphic and structural reappraisal of Central Malawi: results of a geotraverse. *Journal of African Earth Sciences*, **8**, 79–90.

PIQUE, A., DAHMANI, M., JEANETTE, D. & BAHI, L. 1987. Permanence of structural lines in Morocco from Precambrian to Present. *Journal of African Earth Sciences*, **6**, 247–256.

PORADA, H. 1989. Pan-African rifting and orogenesis in southern to equatorial Africa and eastern Brazil. *Precambrian Research*, **44**, 103–136.

REEVES, C. V. 1978. Final interpretation report. Reconnaissance Aeromagnetic Survey of Botswana 1975–1977. Canadian International Development Agency. Government Printer, Gaborone.

SACCHI, R., MARQUES, J., COSTA, M. & CASATI, C. 1984. Kibaran events in the southernmost Mozambique Belt. *Precambrian Research*, **25**, 141–159.

SANDERS, L. D. 1965. Geology of the contact between the Nyanza Shield and the Mozambique Belt in western Kenya. *Bulletin Geological Survey Kenya*, **7**.

SHACKELTON, R. M. 1986. Precambrian collision tectonics in Africa. *In*: COWARD, M. P. & RIES, A. C. (eds) *Collision Tectonics*. Geological Society, London, Special Publication, **19**, 329–352.

STOWE, C. W. 1974. Alpine-type structures in the Rhodesian basement complex at Selukwe. *Journal of the Geological Society, London*, **130**, 411–425.

——, HARTNADY, C. J. H. & JOUBERT, P. 1984. Proterozoic tectonic provinces of southern Africa. *Precambrian Research*, **25**, 229–231.

SUTTON, J. & WATSON, J. V. 1974. Tectonic evolution of continents in early Proterozoic times. *Nature*, **247**, 433–435.

TANKARD, A. J., JACKSON, M. P. A., ERIKSSON, K. A., HOBDAY, D. K., HUNTER, D. R. & MINTER, W. E. L. 1982. *Crustal Evolution of Southern Africa*. Springer, New York.

TAYLOR, P. N., MOORBATH, S., LEUBE, A. & HIRDES, W. 1988. Geochronology and crustal evolution of early Proterozoic Granite-Greenstone Terraines in Ghana/West Africa. Abstract Volume. *International Conference on the geology of Ghana, 9–16 October, 1988*, 43.

TRELOAR, P. J. 1988. The geological evolution of the Magondi Mobile Belt, Zimbabwe. *Precambrian Research*, **38**, 55–73.

VAN HOUTEN, F. B. & HARGRAVES, R. B. 1987. Palaeozoic drift of Gondwana: Palaeomagnetic and stratigraphic constraints. *Geological Journal*, **22**, 341–359.

VICAT, J. P. & VELLUTINI, P. J. 1988. Geologie et geochimie de la serie precambrienne de la Bikossi, le long du realignment du chemin de fer congo-ocean, dans la chaine du Mayambe (Republique populaire du Congo). *Journal of African Earth Sciences*, **7**, 811–820.

VILJOEN, M. J. & VILJOEN, R. P. 1970. Archaean vulcanicity and continental evolution in the Barberton region, Transvaal. *In*: CLIFFORD, T. N. & GASS, I. G. (eds) *African Magmatism and Tectonics*. Oliver & Boyd, Edinburgh, 27–49.

WHITE, R. & MCKENZIE, D. 1989. Magmatism at rift zones: the generation of volcanic continental margins and flood basalts. *Journal of Geophysical Research*, **94**, 7685–7730.

WILLIAMS, H. R. 1978. The Archaean geology of Sierra Leone. *Precambrian Research*, **6**, 251–268.

WINDLEY, B. F. 1985. *The Evolving Continents*. John Wiley, Chichester (2nd edition).

WRIGHT, J. B., HASTINGS, D. A., JONES, W. B. & WILLIAMS, H. R. 1985. *Geology and Mineral Resources of West Africa*. George Allen & Unwin, London.

From WRIGHT, E. P. & BURGESS, W. G. (eds), 1992, *Hydrogeology of Crystalline Basement Aquifers in Africa*
Geological Society Special Publication No 66, pp 59–76.

The hydrology of basement complex regions of Africa with particular reference to southern Africa

F. A. K. Farquharson & A. Bullock

Institute of Hydrology, Maclean Building, Wallingford, Oxon, OX10 8BB, UK

Abstract. Basement complex rocks occur extensively throughout Africa and experience a wide range of climatic conditions. Over much of the continent, potential evaporation exceeds rainfall and runoff is consequently limited. Much of the continent is short of reliable water supplies, and hence understanding the hydrological behaviour of river basins throughout this important region of Africa is of vital importance to the development of the area.

There is a considerable body of rainfall and river flow data throughout the continent of Africa, but whilst climatic data has been collated and analysed on the regional or continental scale, river flow data has been analysed only at the national or sub-regional scale.

An attempt is made here to summarize the hydrological characteristics of those regions of Africa having outcrops of basement complex rocks, illustrating general observations with the results of detailed analyses of available data from Malawi and Zimbabwe. It is suggested that the results of studies in these two countries may be of wider benefit to workers in southern and eastern Africa in particular.

As far as the hydrology of African basement aquifers is concerned, the climatic variables of rainfall and potential evaporation are the most significant factors in determining runoff and groundwater recharge. Both are affected by the seasonal movements of the sun; the pattern of seasonal rainfall in particular being derived from the movement of large-scale zones of high and low pressure caused by the migration of the sun between the tropics.

The majority of the region of interest is formed by a high plateau of relatively uniform elevation with no significant mountain ranges to influence the movement of air masses, except for a few examples such as the Ethiopian highlands and the Fouta Djalon in Guinea. Thus the seasonal system of winds and rainfall depends mainly on the general circulation patterns of the atmosphere, which are driven by the seasonal migration of the sun between the tropics. The zone of greatest incoming solar radiation moves from the southern hemisphere in January to the Tropic of Cancer in the north during July, followed by a trough of low pressure with a lag of about one month. Moisture is brought to the continent through the inflow of maritime air masses from the Indian and Atlantic Oceans, and for much of those regions of Africa covered by basement complex rocks (Fig. 1), the easterly air streams from the Indian Ocean and the southwesterly winds from the southern Atlantic are of primary importance.

A simplified diagram of seasonal pressure and air flow patterns is given in Fig. 2. This shows how the trough of low pressure produces a zone of converging air from the higher pressure zones on either side. This is the Intertropical Convergence Zone, or ITCZ, and is responsible for much of the seasonal variation in rainfall throughout tropical and sub-tropical Africa.

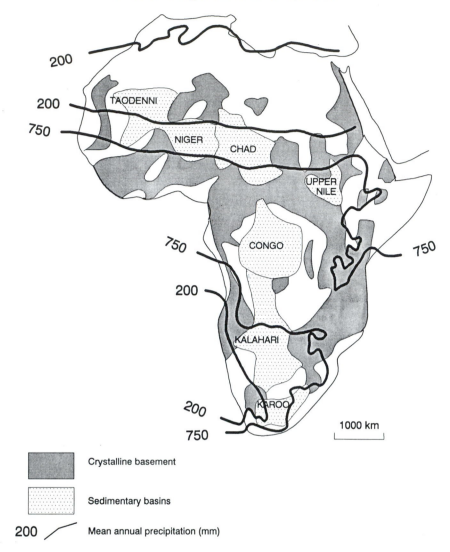

Fig. 1. Generalized outcrop of crystalline basement rocks in Africa.

In January, southern Africa is affected by onshore, moisture bearing winds, and this region consequently has summer rainfall. In July, with the sun moving in a northerly direction, the interior of southern Africa becomes cooler than the Indian Ocean, causing the development of a continental high-pressure system. The descending and outflowing air leads to the southern dry season.

The northeastern region is affected to a limited extent by an Asian northeast monsoon, and many parts of Somalia, Kenya and Ethiopia also receive rainfall during the months of November to January. However, seasonal rainfall in East Africa is very complex and spatially variable, and is difficult to summarize. Most of Tanzania has the pronounced summer rainfall maximum typical of southern Africa, but northwestern Tanzania, much of Kenya and southern Somalia and Ethiopia display a bi-modal seasonal rainfall pattern, as shown in Figs 3 & 4. Similarly, parts of West Africa have a bi-modal rainfall pattern caused by the north to south migration of the ITCZ.

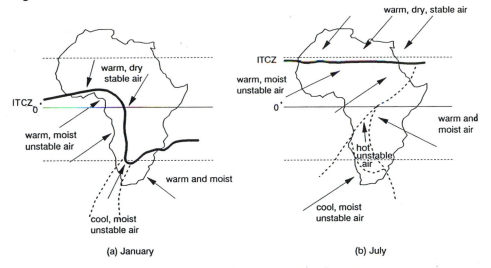

(a) January   (b) July

Fig. 2. Mean position of the Intertropical Convergence Zone (a) in January, (b) in July (from Balek, 1977).

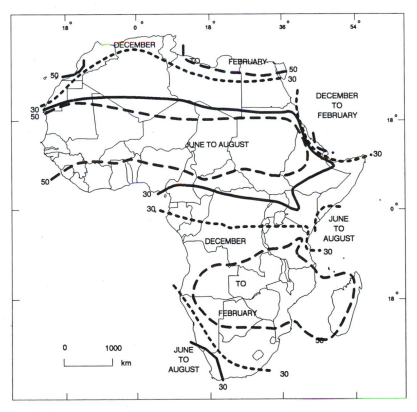

Fig. 3. Period of wet season, and proportion of annual rainfall falling during the wet season, given as percentage (from Griffiths 1983).

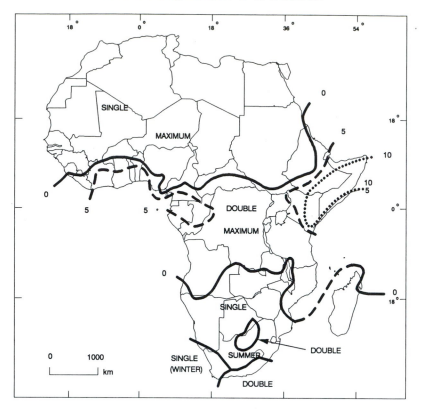

Fig. 4. Nature of annual rainfall distribution. Values of

$$\frac{\text{secondary maximum} - \text{secondary minimum}}{\text{annual total}}$$

given as a percentage (from Griffiths 1983).

This double rainfall season has both beneficial and detrimental effects on water resources and agriculture throughout the region. Whilst it may be possible to produce two crops in parts of this region because of the two inputs of rainfall, runoff is generally lower for any given annual average rainfall than would be the case for a single rainfall zone. This is because soil moisture deficits must be made up before runoff can occur, and this must be achieved twice in the year in regions having a bi-modal rainfall distribution. Similarly, greater evaporation losses ensue during two rainfall seasons than during a single summer peaked season, where for most of the year the soil is dry and actual evaporation is very low. Thus given an annual rainfall of say 1000 mm, runoff and water resources potential will be greater in the single rainfall season zones. The distinction between potential and actual evaporation will be expanded later.

The variability of seasonal rainfall and the spatial variability of rainfall was illustrated by Sutcliffe & Knott (1987) who considered rainfall along a south to north transect from Abidjan at 5° N to Timbouctou at 16° N in West Africa. A diagram from their paper, reproduced here as Fig. 5(a), shows how annual average rainfall decreases from 2000 mm near the coast to 200 mm some 1300 km to the north. The seasonal distribution of this rainfall is shown in Fig. 5(b), where the changing pattern in response to the migration of the ITCZ is apparent.

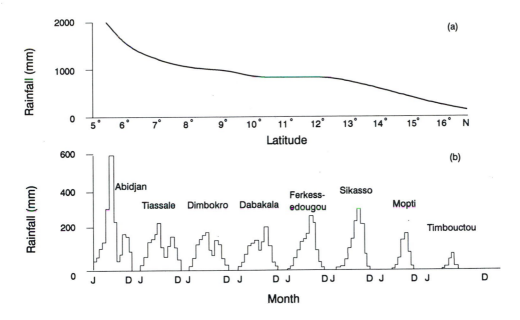

Fig. 5. Mean annual rainfall (a) and seasonal distribution of rainfall (b) for a transect of latitude in West Africa (from Sutcliffe & Knott 1987).

Evaporation

Evaporation is the second major component of the hydrological balance of the area of interest, as in simple terms, runoff is the difference between gross rainfall and actual evaporation. Much of the observed outcrop of basement complex rocks occurs between the tropics (Fig. 1) and receives high inputs of solar radiation. Because of this, potential evaporation throughout the region is generally high, ranging from over 3000 mm in the arid regions of Sudan to perhaps 1300 to 1400 mm in highland regions of Malawi, Rwanda and Burundi (Woodhead, 1968; Balek, 1977; Mandeville & Batchelor, 1990).

The influence of altitude on evaporation is evident in Malawi, where higher values of potential evaporation are recorded along the Rift Valley floor (up to 1840 mm) and lower values are found on the upland areas, with a minimum of 1316 mm. Data from Malawi (Mandeville & Batchelor 1990) indicate that there is very little variation amongst annual values of potential evaporation at individual meteorological stations, with coefficients of variation of 5% or below.

Actual annual evaporation losses as well as seasonal distribution are closely related to average annual rainfall. Actual losses calculated from simple water balance for 102 catchments in Zimbabwe, Malawi and Tanzania (Bullock, Chirwa, Matondo & Mazvimavi 1990), predominantly on basement aquifers, exhibit a positive relationship with rainfall, when expressed as a ratio of actual to potential losses (Fig. 6). On the whole there is a consistent relationship between the ratio, r, and annual rainfall, but the one or two very low values, and particularly the single high observed value of 1.2 may indicate poor data. Typical values of the ratio (r) of actual to potential evaporation for values of average annual rainfall are summarized in Table 1. Actual losses are less than 25% of potential losses in regions of Tanzania and Zimbabwe, and values exceed 60% of potential losses in only 3% of catchments analysed, being the catchments receiving the highest rainfall totals.

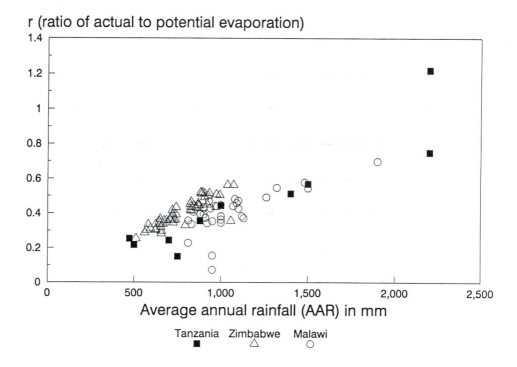

Fig. 6. Ratio of actual to potential evaporation (r) in Tanzania, Zimbabwe and Malawi.

Table 1. *Ratio (r) of actual to potential annual evaporation for a range of average annual rainfalls*

			Average annual rainfall			
500 mm	750 mm	1000 mm	1250 mm	1500 mm	1750 mm	2000 mm
0.25	0.35	0.45	0.5	0.55	0.65	0.75

Runoff

The runoff process is largely controlled by two factors. The first of these is rainfall, particularly the spatially variable nature of the convective storms that characterize the rainfall process throughout the area, and in annual terms, whether the rainfall occurs in one or two wet seasons. The second factor is the local soil profile, which affects the infiltration of rainfall into the soil, and vertical deep recharge and lateral flow downslope of this water. The soil is itself affected by the underlying geology and by variations in the local topography.

It is useful to consider the effects of runoff of the soil profile firstly, because understanding this mechanism helps to explain how the rainfall process affects runoff. Wright (this volume) describes the typical soil profile developed by weathering of basement complex rocks and suggests that the predominant soil types derived from this weathering are sandy soils on the interfluves and clay soils on the valley bottoms. These valley clays are in part colluvial in origin, being washed down from the upper slopes, but are predominantly secondary clays, such as kaolinite, derived in situ through chemical action, which is most active in the vadose layer and zone of water table fluctuation. This leads to the development of so-called dambos or vleis in low-lying areas. These are areas of low relief with poorly drained clay soils which are able to store significant quantities of water.

The overall effect of this weathering process is a catena of soils down the valley, from permeable sandy soils on the interfluves, through finer sands with colluvial clay reducing permeability on the valley sides, to poorly drained clay soils in the valley bottoms. Rainfall will tend to pass quickly into the upper sandy soils, but reaches layers of reduced permeability at relatively shallow depths, often at the stone line and at the top of the weathered saprolite layer. Thus runoff is initially vertical, but at shallow depths the reducing permeability produces lateral throughflow downslope towards the valley bottom dambos. Surface runoff rarely occurs except in localized low-lying areas where the soil profile may become saturated. Throughout the area, some of the infiltrating water also passes down through zones of localized higher vertical permeability to deeper groundwater zones to sustain springflows and river baseflows.

On an annual basis, the runoff process is strongly influenced by the seasonal distribution of rainfall and potential evaporation as described earlier. Because there is little, if any, rainfall during some months of the year, a significant soil moisture deficit develops during the dry season, when potential evaporation exceeds rainfall. At the onset of the rainy season, much of the rainfall goes towards making good this deficit and little initial runoff occurs. Where there are two dry seasons, there will in general be a longer period during which transpiration from vegetation can take place and there will be two occasions on which this soil moisture deficit must be replenished. Therefore runoff from areas having a bi-modal rainfall distribution is lower than for the same gross rainfall occurring in a single season. Discussion on how single and double wet seasons affect runoff is given in Sutcliffe & Piper (1986), where the typical seasonal pattern of runoff is illustrated in Fig. 7, which shows the major water balance components for a station in West Africa.

Runoff from catchments is the residual of rainfall minus actual evaporation losses. Observed catchment runoff:rainfall ratios on basement aquifers are variable in the same way as actual losses, and for example range between 4% to 54% amongst 38 catchments in Malawi (Hill & Kidd 1980) and from 2% to 50% amongst 108 catchments in central Zimbabwe. Higher catchment ratios are normally associated with higher average annual rainfall depths reflecting the basic linear nature of regional rainfall:runoff relationships. This point is illustrated in Fig. 8, which is typical of what one often observes throughout Africa; the relationship between

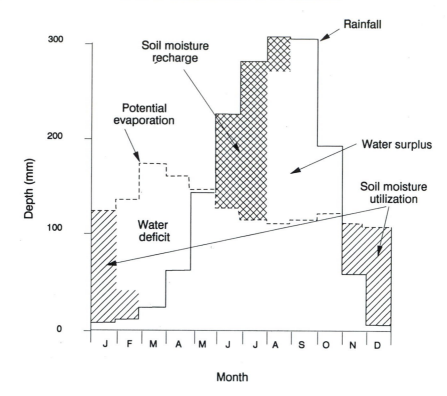

Fig. 7. Schematic diagram of annual water balance components in West Africa (from Sutcliffe & Piper 1986).

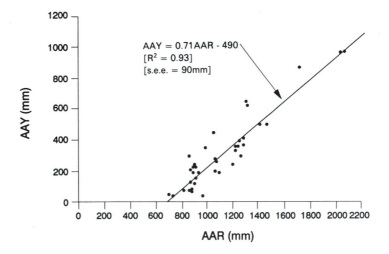

Fig. 8. Relationships between average annual yield, AAY (mm), and rainfall, AAR (mm) for 38 catchments in Malawi (from Drayton *et al.* 1980).

annual runoff and annual rainfall is often linear and has a finite intercept on the rainfall axis implying little or no annual runoff until some threshold rainfall has been reached. This threshold is the dry season soil moisture deficit discussed above. Mean ratios from the two regional studies are 27% in Malawi and 17% in Zimbabwe, with both areas located within the Indian Ocean Slope zone (UNESCO 1978), for which the mean runoff ratio is 11%.

Unlike potential evaporation, variability of annual runoff totals is high, and data from southern Africa confirm the conclusion of MacMahon et al. (1987) that there is greater variability of annual runoff in regions of low runoff, although they note that Australia and southern Africa have a higher coefficient of variation of annual runoff than other continents. Malawian catchments, with annual runoff totals in excess of 200 mm, tend to have coefficients of variation of annual totals of less than 60%. In contrast, Zimbabwe catchments with annual runoff totals less than 200 mm, exhibit coefficients of variation in excess of 60% and as high as 120%.

Resources and recharge

Resource assessment must be based on the analysis of hydrometeorological data, whose coverage varies throughout Africa as illustrated in Table 2.

Of the top ten countries (excluding island nations) with highest densities of flow gauging stations in Africa, six (Kenya, Malawi, Burundi, Lesotho, Zimbabwe and South Africa) are in southern or eastern Africa. Only two countries in West Africa (Togo and Gambia) have more than one gauging station per 2000 km^2. Rain gauge densities reflect a similar pattern. However, even amongst those countries with the higher densities of hydrometeorological measurement, the network does not allow detailed understanding of small-scale or meso-scale variability, except in those zones within countries where there are clusters of observation. In general, there is an absence of small gauged catchments, with the majority tending to be in excess of 100 km^2 or more, and the dominance of non-recording discharge stations reduces the quality of data resolution on a temporal scale.

Two potential sources of water are available: surface runoff from rivers and groundwater via dug wells or boreholes. Surface water resources have the limitation of being strongly seasonal in response to the seasonal rainfall characteristic of most of the area under consideration and hence storage is needed to ensure all year supply. For an individual family group, domestic water supplies may be met by storage of runoff from roofs, although the quality of such water may deteriorate during the prolonged dry season. For larger village or town requirements, a larger surface water storage in the form of reservoir may be required, particularly if industrial or animal watering demands are also to be met.

Reservoirs are expensive to construct and maintain and good storage sites are limited because of the low relief typical of the weathered basement complex. The low surface runoff typical of this geology means that deeply incised valleys suitable for dam construction are rarely produced. Such reservoirs as do exist tend to have a shallow saucer-like form with a large surface area for any given storage capacity. This large surface area for any particular storage combined with the high potential evaporation typical of the region means that reservoirs typically lose as much water through evaporation as they yield to water supply. Thus, in order to obtain a particular yield from a reservoir with the flat shape possible on basement complex

Table 2. *Availability of rainfall, evaporation and river flow data in different regions of Africa*

Country	Rain gauges		Evaporation pan		Flow gauges	
Algeria	970	0.41	45	0.02	0	0
Angola	610	0.49	30	0.02	128	0.10
Benin	79	0.70	17	0.15	30	0.27
Botswana	61	0.10	2	0.01	44	0.08
Burkina Faso	198	0.72	13	0.05	49	0.18
Burundi	119	4.25	3	0.11	32	1.14
Cameroon	193	0.41	10	0.02	81	0.17
Cape Verde	0	0	0	0	0	
C. African Republic	94	0.15	5	0.01	68	0.11
Chad	112	0.09	10	0.01	37	0.03
Comoros	58	29.0	1	0.50	0	0
Congo	122	0.36	4	0.01	62	0.18
Cote d'Ivoire	0	0	0	0	0	0
Djibouti	24	1.1	4	0.18	2	0.09
Egypt	82	0.08	15	0.01	230	0.23
Equatorial Guinea	0	0	0	0	0	0
Ethiopia	482	0.39	60	0.05	271	0.22
Frengh Reunion	152	60.8	3	1.2	15	6.0
Gabon	70	0.26	2	0.01	43	0.16
Gambia	20	1.82	1	0.09	7	0.64
Ghana	526	2.20	51	0.21	116	0.49
Guinea	49	0.20	0	0	34	0.14
Guinea Bissau	29	0.81	1	0.03	0	0
Kenya	1282	2.20	136	0.23	838	1.44
Lesotho	46	1.53	9	0.30	34	1.13
Liberia	31	0.28	0	0	11	0.10
Libya	148	0.08	18	0.01	30	0.02
Madagascar	586	1.00	8	0.01	100	0.17
Malawi	719	6.09	60	0.51	165	1.40
Mali	184	0.15	13	0.01	28	0.02
Mauritania	39	0.04				
Mauritius	319	159.5	11	5.5	65	32.5
Morocco	279	0.64	59	0.13	144	0.32
Mozambique	940	1.20	36	0.05	192	0.25
Namibia						
Niger	111	0.09	10	0.01	36	0.03
Nigeria	1100	1.19	63	0.07	203	0.22
Portugal – Madeira	2	2.5	0	0	0	0
Rwanda	2	0.08	7	0.27	17	0.65
Sao Tome and Principe	194	201.0				
Senegal	116	0.59	29	0.15		
Seychelles	43	154.7	4	14.4	21	75.5
Sierra Leone						
Somalia			8	0.01		
South Africa	1096	0.90	202	0.17	821	0.67
Spain – Canary Islands	256	35.1	1	0.14	0	0
Sudan	1264	0.50	28	0.01	236	0.09
Swaziland						
Togo	102	1.82	9	0.16	42	0.75
Tunisia	750	4.6	18	0.11	197	1.20

The header spans: "Absolute number of gauges followed by number per 1000 km²"

Table 2. *(continued)*

Country	Rain gauges		Evaporation pan		Flow gauges	
Uganda	613	2.60	38	0.16	73	0.31
United Rep. of Tanzania	960	1.02	64	0.07	174	0.18
Zaire	272	0.12	22	0.01	10	0.01
Zambia	775	1.03	27	0.04	154	0.20
Zimbabwe	2462	6.30	108	0.28	340	0.87

Header note: Absolute number of gauges followed by number per 1000 km^2

Source: Infohydro, WMO, 1987.

The authors believe that minor discrepancies may exist in the table and that, for example, the entry for Cote d'Ivoire indicates an incomplete set of data in the table rather than a total lack of hydrometeorological data in this country. Readers will no doubt be aware of other discrepancies.

catchments, the reservoir must have a capacity of up to twice that which would be required without the high evaporative loss.

In general, surface water sources are only viable within the region for urban or industrial purposes, where high water tariffs can be charged to customers. Village supplies, to largely agricultural communities, can rarely be provided economically from reservoirs due to the high cost of the construction of dam and treatment works. Surface water supplies are almost never viable for irrigated crops due to the limited markets for produce. Perhaps in the long term, irrigation might be feasible in isolated cases close to urban areas which could provide a market for high-value cash crops. Alternatively, if good quality cash crops could be exported, the cost of irrigation from surface supplies might be justified. However, the value of small-scale irrigation from groundwater to produce crops for the use of local communities has been clearly demonstrated (Faulkner & Lambert 1991).

Hydrological studies in Zimbabwe and Malawi

Hydrological studies were undertaken as a component of the Basement Aquifer Project (British Geological Survey and others, 1989) to analyse the climatic and geomorphological controls upon hydrological response from 26 catchments in Malawi and Zimbabwe. Conclusions from this study presented here focus upon the quantification of the groundwater baseflow component, and upon climatic and geomorphological factors affecting baseflow:rainfall ratios, total evaporation losses and dry season flows. Details of the gauging stations are presented in Table 3. Catchments were selected to represent a diversity of terrain features, including erosion surfaces, relief and drainage network characteristics, including dambos, while ensuring that available flow records are long, relatively natural and of good quality. Catchment characteristics comprising indices of relative relief, dambo area and perimeter, drainage density, rainfall and potential evaporation were calculated above each gauging station. Flow data were analysed to derive total runoff, total evaporation losses, groundwater baseflow volumes, and low flow statistics. Discussion here focuses upon the conclusions from an examination of the relationships between catchment characteristics and groundwater baseflow and dry season flows.

Table 3. *Gauging station details*

Station No	River name	Period of record	Years of data
1.R.3	Rivi-Rivi	1959/60–1986/87	19
2.B.22	Thondwe	1959/60–1974/75	16
3.E.3	Livulezi	1957/58–1974/75	18
4.B.3	Linthipe	1957/58–1974/75	18
4.B.4	Diampwe	1957/58–1974/75	18
4.D.4	Lilongwe	1955/56–1974/75	20
5.C.1	Bua	1974/75–1986/87	11
5.D.1	Bua	1959/60–1974/75	16
5.D.2	Bua	1953/54–1974/75	22
5.B.13	Kaombe	1976/77–1985/86	8
6.F.2	Luweya	1960/61–1974/75	15
6.F.5	Luchelemu	1958/59–1974/75	17
6.F.10	Luchelemu	1959/60–1974/75	16
7.F.2	Chilinda	1959/60–1974/75	16
7.A.3	South Rukuru	1956/57–1974/75	19
7.A.9	South Rukuru	1972/73–1984/85	13
A12	Tegwani	1951/52–1983/84	26
A55	Tegwani	1969/70–1971/72	3
C33	Sebakwe	1956/57–1983/84	20
C41	Umvumi	1963/64–1983/84	21
C47	Sebakwe	1962/63–1983/84	20
D28	Mazoe	1932/33–1984/85	52
D47	Nyagadzi	1979/80–1984/85	5
D55	Dora	1970/71–1984/85	13
E45	Umtilikwe	1959/60–1985/86	25
E49	Poptekwe	1959/60–1984/85	26

Groundwater baseflow

Hydrograph separation by the Base Flow Index technique (Institute of Hydrology 1980) was applied to time series of daily mean flow data for each of the 26 catchments to quantify groundwater baseflow contributions to streamflow. Calculated annual groundwater baseflow values as absolute depths are mapped in Fig. 9. Observed groundwater baseflow depths range between 0 mm and 370.6 mm, and are strongly controlled by mean annual rainfall, as illustrated by Fig. 10.

Ratios of groundwater baseflow to mean annual rainfall vary from zero to 0.25. Ratio values are generally higher in Malawi, with a range of 0.02 to 0.27 compared to Zimbabwe, where the range is between zero and 0.09. Correlations between the baseflow:rainfall ratio, termed BF/AAR, demonstrate significant positive relationships at the 5% level with mean annual rainfall ($r = 0.843$), mean relative relief ($r = 0.894$) and a negative relationship with standardized dambo perimeter ($r = 0.390$). Mean relative relief is defined as the mean value of relative relief (in metres) calculated for each square kilometre and the standardized dambo perimeter is ratio of dambo perimeter length, in kilometres, to dambo area in square kilometres.

To eliminate non-linearity effects attributable to regional rainfall variations, baseflow contributions to streamflow were standardized under a scenario of 1000 mm

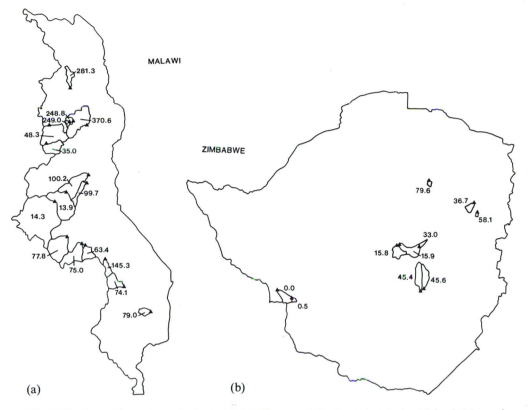

Fig. 9. Study catchments and observed baseflow contributions (mm), in Malawi (a), and Zimbabwe (b).

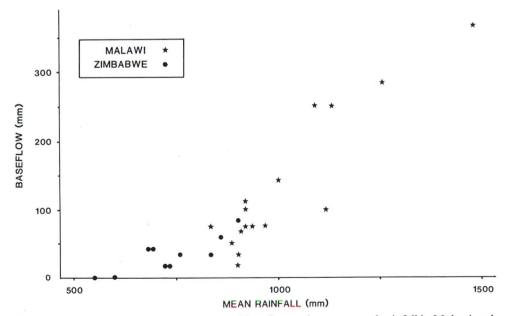

Fig. 10. Relationships between mean annual baseflow and mean annual rainfall in Malawi and Zimbabwe.

annual rainfall, termed BF/AARsim, using catchment regression relationships based upon annual values of rainfall and baseflow. The aim was to reduce the inter-correlation between mean annual rainfall and mean relative relief by deducing through linear regression on gauged catchments the baseflow that would correspond to an average annual rainfall of 1000 m. Under this scenario, BF/AARsim values range from 0.002 to 0.206, with Malawi values ranging from 0.048 to 0.206 and Zimbabwe values generally increase from the observed values to a range between 0.002 and 0.129. Correlations of BF/AARsim against terrain and geomorphological indices demonstrate that the positive trend with relative relief is the only significant relationship at the 5% level (Fig. 11).

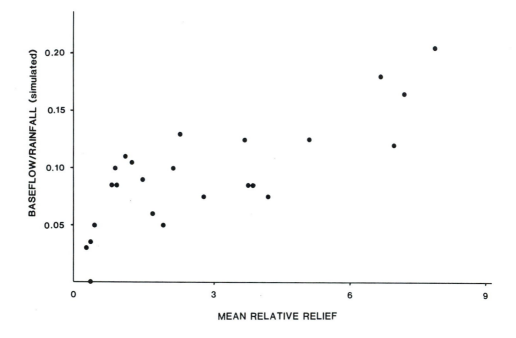

Fig. 11. Relationships between baseflow:rainfall ratio (simulated for an annual rainfall of 1000 mm) and mean relative relief.

It can be concluded that groundwater baseflow as estimated from hydrograph separation ranges in the study catchments between zero and 27% of mean annual rainfall, or between zero and 371 mm in absolute terms. There is a strong rainfall control upon the baseflow:rainfall ratio, but when the effect is removed by standardi-zation, then groundwater baseflow contributions to streamflow appear to be related to relative relief. Catchments with high relative relief, characteristic of the Post-African erosion surface in Zimbabwe and the high residuals along the Malawi Rift shoulder exhibit BF/AARsim ratios greater than 7% and up to 20%. Catchments with low mean relative relief, characteristic of the African surface (typically below 30 metres per km^2) exhibit BF/AARsim ratios of less than 12% and down to zero.

Dry season flows

Flow data series were analysed using standard low flow procedures (Institute of Hydrology 1980) to derive indices of duration of low flows, low flow frequency and recession in the 26 catchments. Investigation of the factors which determine dry season flow regimes reveal a dominant rainfall control, reflecting the reported relationship between groundwater baseflow and rainfall. In general, an insufficiently large data set restricted the identification of strong terrain and geomorphological controls upon dry season flow regimes, specifically because of the significant positive correlations between indices of relative relief and rainfall. Low ranges amongst the observed low flow statistics and absence of clustering amongst catchments within a small rainfall range do not enable conclusions to be drawn concerning the duration and volume of dry season flows. However, more detailed analysis of the low flow data set assembled by Drayton *et al.* (1980) for Malawi (Fig. 12) demonstrates that within a range of mean annual rainfall between 800 mm and 1000 mm there is a tendency for catchments with higher dambo densities (proportion of the catchment defined as dambo) to possess lower Q75(10) values and to cluster below those with lower dambo density. Q75(10) is the 10-day flow that is exceeded for 75% of the time.

In reviewing past studies of dambo hydrology, Bullock (in press) presents an argument for low flow reduction by dambos, challenging the widely-held belief that dambos release water from storage to augment dry season flows. This evidence from Malawi further establishes the argument that dambos act to reduce dry season low flows rather than augment their volume and duration.

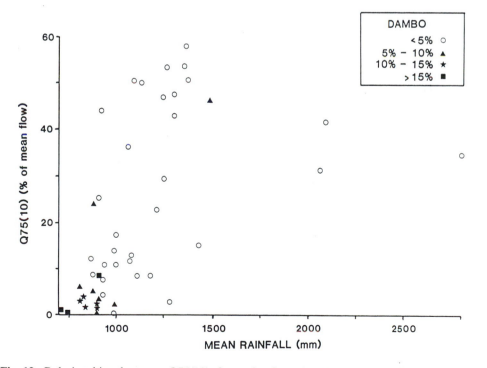

Fig. 12. Relationships between Q75(10) from the flow duration curve and mean annual rainfall with catchments grouped by percentage dambo.

Estimation of groundwater recharge

A number of papers within this volume cover the subject of groundwater resources from basement complex aquifers (e.g. Wright, this volume; McFarlane *et al.* this volume) and hence little comment need be given here on the hydrogeological aspects of groundwater development. However, it may be unwise to invest heavily in development of such resources without assessment of groundwater recharge, and hydrology can provide one method of determining this vital component of the groundwater balance.

There are two main mechanisms of recharge: direct recharge through general infiltration over wide areas, and indirect recharge from concentrations of surface runoff along water courses. Whilst the first mechanism may be significant in areas of higher rainfall, over much of Africa it is probable that the dominant recharge route is from linear concentrations of runoff along valley bottom water courses particularly during isolated spate runoff events following high intensity convective storm rainfall. Where annual rainfall is less than 400 mm, Edmunds *et al.* (1988) showed that recharge from direct infiltration is likely to be small or negligible and hence renewable groundwater resources would sustain only a sparse pastoral population through dug wells or shallow boreholes.

Little research has been undertaken into indirect recharge from runoff into river alluvium or deeper regolith sequences. However, in many parts of southern Africa, boreholes into the alluvial bed of sand rivers provide a sustainable source of water even where the annual average rainfall may be no more than 500 mm. There has even been some work on developing small-scale irrigation from sand rivers in Botswana and Zimbabwe, although the authors have no details of the success of such schemes. The advantage of developing such resources is that once runoff is concentrated underground in the alluvial river beds, evaporation losses are small. The occasional spate runoff events typical of these arid and semi-arid regions prevent vegetation from becoming established in the water courses, and hence there is no scope for transpiration of such water. Even where the water table is close to the surface in sand rivers, evaporation losses are small as diffusion of water vapour upward through the sands is a very inefficient and slow process and because capillary rise through the predominantly coarse alluvium is negligible.

Estimation of groundwater recharge through chloride balances of rainfall and groundwater and by the analysis of water level changes in boreholes and wells are discussed elsewhere in this volume. The following section, however, discusses the analysis of gauged streamflow data to estimate groundwater recharge in Zimbabwe and Malawi.

Analysis of hydrological data has focused upon quantification of groundwater baseflow and dry season flows. Conclusions from these studies can be brought together to develop estimates of groundwater recharge within the gauged catchments and to consider the controlling factors upon recharge.

Observed groundwater baseflow contributions to streamflow, expressed as a fraction of rainfall, range from zero to 25% and provide minimum estimates of groundwater recharge. Groundwater baseflow contributions to streamflow are strongly controlled by annual rainfall depths, and, when standardized by rainfall, appear to relate to relative relief. Catchments characteristic of the Post-African surface of Zimbabwe and the Rift Valley shoulder of Malawi exhibit baseflow:

rainfall ratios between 0.02 and 0.20 while ratios on the African surface are typically less than 0.12 and can be as low as zero in western Zimbabwe.

Groundwater baseflow estimates from this study should be considered as a minimum because recharged water can be lost to evaporation through a number of components of the hydrological cycle before being measured at a gauging station. Greater emphasis should be given to conclusions from process studies of evaporation to estimate the loss of recharged water to evaporation, such as those methods involving the regional extrapolation of point measurements of evaporation (British Geological Survey and others 1989).

Conclusions

Basement aquifers are widespread throughout Africa, as shown in Fig. 1, and occur under a wide range of climatic conditions. Despite a relative wealth of hydro-meteorological data, there has been little effort amongst researchers to characterize the spatial and temporal variability of runoff and recharge from such regions. The emphasis has been on the analysis of data at a local regional or national scale, often due to the difficulty of assembling sufficient good quality data at the broader regional scale. This lack of general areal analysis of the data should be contrasted with climatology, where a continental picture has been developed from all available data. There is a need for a similar widespread analysis of all available hydrological data, if such data could be more generally available to researchers and if it could be assembled onto a common database. Unfortunately, there is at present no readily available source of the required data at the regional or continental scale; an omission which needs to be addressed by one of the international agencies such as the U.N., perhaps through the World Meteorological Organisation.

Nevertheless, research has been undertaken at the national or sub-regional level, and has demonstrated the significant variability in water balance components affecting this important aquifer.

Hydrological studies were undertaken as a component of the Basement Aquifer project (British Geological Survey and others, 1989) to analyse the climatic and geomorphological controls upon hydrological response from 26 catchments in Malawi and Zimbabwe. Conclusions from this study presented here focus upon the quantification of the groundwater baseflow component, and upon climatic and geomorphological factors affecting baseflow:rainfall ratios, total evaporation losses and dry season flows.

Observed groundwater baseflow contributions to streamflow, expressed as a fraction of rainfall, range from zero to 25% and provide minimum estimates of groundwater recharge. Groundwater baseflow contributions to streamflow are strongly controlled by annual rainfall depths, and, when standardized by rainfall, appear to relate to relative relief. Catchments characteristic of the Post-African surface of Zimbabwe and the Rift Valley shoulder of Malawi exhibit baseflow: rainfall ratios between 0.02 and 0.20 while ratios on the African surface are typically less than 0.12 and can be as low as zero in western Zimbabwe.

There is also a lack of research into hydrological processes throughout the region, which is again hindering the development of the water resources of those regions of Africa covered by basement complex rocks. In particular, few studies have been undertaken to monitor runoff from small catchments, thus making it difficult to estimate the proportion of rainfall contributing to runoff. Whilst there are many large-gauged catchments with areas of hundreds or thousands of km², as shown in Table 2, because of the spatial variability of rainfall and the generally limited raingauge coverage, it is not possible to quantify runoff in many cases. Thus, is an observed runoff of 10 per cent caused by a uniform runoff over the whole catchment or by 20 per cent over only half of the catchment? Without data from small catchments, or without better instrumented larger catchments, reliable estimates of the runoff coefficient cannot be developed.

Finally, for many purposes it is likely that groundwater will provide a more reliable source of water than surface water. Rivers are in most cases seasonal, and the basement complex rocks do not in general weather to produce incised valleys suitable for reservoir development. However, in Botswana surface water has been shown to be the most economical source of supply for industrial and city demands despite the poor natural damsites available.

References

BALEK, J. 1977. *Hydrology and Water Resources in Tropical Africa*. Elsevier, Amsterdam.
BRITISH GEOLOGICAL SURVEY and others 1989. *The Basement Aquifer Research Project 1984– 1989: Final Report to the Overseas Development Administration*. Technical Report WD/89/15.
BULLOCK, A. in press. Dambo Hydrology in southern Africa: Review and reassessment. *Journal of Hydrology*.
——, CHIRWA, A. B., MATONDO, J. I. & MAZVIMAVI, D. 1990. Analysis of flow regimes in Malawi, Tanzania and Zimbabwe: A feasibility study for Africa FRIEND.
DRAYTON, R. S., KIDD, C. H. R., MANDEVILLE, A. N. & MILLER, J. B. 1980. A regional analysis of river floods and low flows in Malawi. Institute of Hydrology Report No. 72, Wallingford, U.K.
EDMUNDS, W. M., DARLING, W. G. & KINNIBURGH, D. G. 1988. Solute profile techniques for recharge estimation in semi-arid and arid terrain. *In*: I. SIMMERS (ed.) *Estimation of Natural Groundwater Recharge*. Reidel, Dordrecht, 139–157.
FAULKNER, R. D. & LAMBERT, R. A. 1991. The effects of irrigation on dambo hydrology: A case study. *Journal of Hydrology*, **123**, 147–161.
GRIFFITHS, J. F. (ed.), 1983. *World Survey of Climatology*, Volume 10, Africa. Elsevier, Amsterdam.
HILL, J. L. & KIDD, C. H. R. 1980. *Rainfall–Runoff Relationships for 47 Malawi Catchments*. Report TP7, Water Resources Branch, Malawi.
INSTITUTE OF HYDROLOGY, 1980. Low Flow Studies Report, Wallingford, UK.
MACMAHON, T. A., FINLAYSON, B. L., HAINES, A. & SRIKANTHAN, R. 1987. Runoff variability: a global perspective. *In: The influence of climate change and variability on the hydrological regime and water resources*. IAHS Publn. No. 168, 3–11.
MANDEVILLE, A. N. & BATCHELOR, C. H. 1990. Estimation of actual evapotranspiration in Malawi. Institute of Hydrology Report No. 110, Wallingford, U.K.
SUTCLIFFE, J. V. & PIPER, B. S. 1986. Bilan hydrologique en Guinee et Togo-Benin, *Hydrologie Continentale*, Vol. I, No. 1, 51–61.
—— & KNOTT, D. G. 1987. Historical variations in African Water Resources. Proceedings of the Vancouver Symposium, August 1987. IAHS Publn. No. 168.
UNESCO, 1978. World water balance and resources of the earth. UNESCO, Paris.
WMO, 1987. InfoHydro Manual, World Meteorological Organisation Operational hydrology Report No. 28, Geneva.
WOODHEAD, T., 1968. *Studies of Potential Evaporation in Kenya*. Ministry of Natural Resources, EAFRO, Kenya.

From WRIGHT, E. P. & BURGESS, W. G. (eds), 1992, *Hydrogeology of Crystalline Basement Aquifers in Africa*
Geological Society Special Publication No 66, pp 77–85.

Structural influences on the occurrence of groundwater in SE Zimbabwe

D. Greenbaum

British Geological Survey, Keyworth, Nottingham, NG12 5GG, UK

Abstract. The Zimbabwe Craton is a heterogeneous assemblage of crystalline basement rocks in which the groundwater reservoirs are structurally controlled or confined to the weathered overburden. Lineaments, interpreted from Landsat imagery and aerial photographs, represent faults, joints and dykes of several ages. Such fissured rocks are more susceptible to deep weathering and are a main target for borehole siting. An analysis of borehole yields indicates that, although failure rates can be high, fractures of all trends are capable of providing an adequate hand-pump yield (0.25 l/s). Thus, despite their origin as compressional shears, many fractures now appear to provide open conduits for groundwater. It is concluded that within the near-surface zone of most importance to basement rock aquifers (40 to 80 m) fractures of all origins are in a state of tension probably as a result of recent uplift and erosional unloading.

The crystalline basement rocks of the cratons possess little primary intergranular porosity or permeability and their hydrogeological properties are thus mainly determined by secondary effects such as brittle fracturing and weathering that give rise to secondary storativity and transmissivity (Clark 1985). In arid areas, weathering etches out lines of structural weakness, such as faults, joints and dykes, which are evident on aerial photographs as 'lineaments' and which provide a focus for more detailed ground geophysical surveys. In the present study, lineaments, interpreted from both Landsat imagery and aerial photographs, were investigated both on the ground and in relation to the hydrogeological database of existing boreholes. Much of this work has already been reported (Greenbaum 1986, 1987, 1988, 1989); this paper summarizes some of the more general conclusions and describes additional correlation studies.

Geological & structural studies in SE Zimbabwe

Geological setting

The Zimbabwe Craton is an Archaean complex of greenstones, mafic and ultramafic rocks, gneisses and migmatites, and late intrusive granites. It is intruded by the Great 'Dyke', overlain by supra-crustal units and margined on at least three sides by mobile belts.

The greenstones are remnants of Archaean volcano-sedimentary piles, now represented as schist belts comprising both volcanic rocks and metasediments. The widespread 'older' gneisses and migmatites consist mainly of tonalites of highly variable composition, with a metamorphic texture and evidence of shearing and recrystallization. They evolved over a long interval together with the greenstones with which they were deformed and which they now enclose and embay. The

'Younger' batholithic granites are generally massive, intrusive rocks of adamellitic and tonalitic composition. The end of the Archaean is marked by emplacement of the Great 'Dyke'. Along the SE margin of the Craton lies the Limpopo Mobile Belt. It consists of high-grade granulite gneisses representing in part the reworked granite–greenstone terrain of the adjacent craton. The main geological subdivisions of southeast Zimbabwe are shown on Fig. 1. For a detailed account of the geology the reader is referred to reports by Macgregor (1947; 1951); Stowe (1980), Bliss & Stidolph (1969), Coward (1976), Coward *et al.* (1976), Wilson (1973; 1979) and Wilson *et al.* (1987) among others.

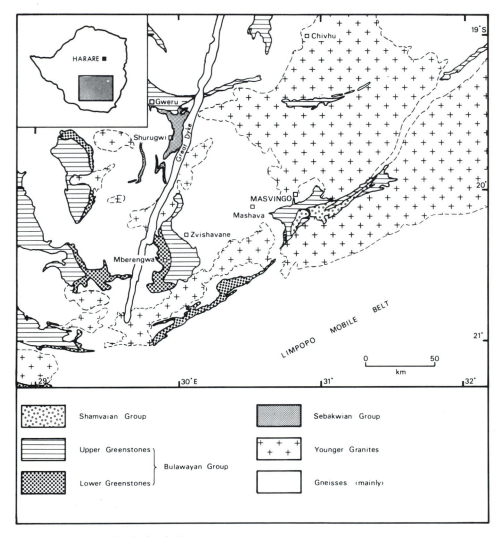

Fig. 1. Principal geological subdivisions of the Masvingo area, SE Zimbabwe.

Remote sensing and field studies

Aerial photographs at scales of 1:25 000 and 1:80 000, and dry season Landsat Multispectral Scanner (MSS) imagery were used in the study. MSS scenes 169-074 and 170-074 were processed on an I^{2}S image analysis system to produce edge-enhanced false colour composites of bands 4-5-7 and a black-and-white image of band 5. The regional fracture pattern was interpreted from both 1:250,000 Landsat images and 1:80 000 aerial photographs and 14 sub-areas were also examined in detail using 1:25 000 scale aerial photographs. The digitized lineament data were analysed to produce summary statistics, rose diagrams, lineament plots (total and selected populations) and lineament density plots.

The ground elements that make up a lineament depend on the scale and resolution of the remote sensing data used. Lineaments interpreted from 1:250,000 satellite images are usually more than 2.5 km long, are negative relief features such as large valleys, and typically represent complex fracture zones. By contrast, air photo lineaments are generally much shorter, correspond to minor drainage features or tonal soil changes and are expressions of individual fracture traces. Dykes may form depressions or ridges or have no relief expression at all, but are often characterized by dark soils, rubbly sub-outcrop and denser vegetation.

Since satellite images and aerial photographs emphasize different aspects of the structure, each can provide complementary information on the hydrogeology of an area. The scale and stereo capability of air photos make them ideal for detailed site selection, for which they are widely used in southern Africa, whereas the synoptic view provided by satellite imagery reveals regional patterns. Furthermore, because open fissure systems support green-leaf vegetation even under drought conditions, dry season satellite imagery in the visible/near infrared can help identify the larger water-bearing fractures. Differences in lineament pattern between satellite and air photo data are common and are illustrated in Fig. 2, which combines length–frequency rose diagrams for the Masvingo area from these two sources. Here, the ENE directions prominent on aerial photographs correspond to foliation-parallel joints in the gneisses; despite their individual small size, and consequent poor expression on satellite imagery, they can form an important source of groundwater locally.

Although lithology is not in itself greatly important in basement hydrogeology, since all crystalline rocks are virtually impermeable in their unweathered state, mineralogy, texture and granularity do affect the ways in which rocks fracture and decompose. Differences in lineament pattern and density are evident from one basement formation to another: for example, fractures that can be traced for several kilometres through the Younger Granites appear to terminate abruptly in the Older Gneiss formation (and may even be expressed as ductile folds within the greenstones). It would seem that isotropic, granular granites fail brittly along few, widely spaced and continuous fractures whereas foliated gneisses dissipate the stresses and yield along multiple, irregular fractures. This in turn has an important effect upon weathering. In arid climates deep weathering tends to be concentrated along fissured zones. Consequently in Zimbabwe the plutonic granites and high-grade granulites form broken, high-relief terrain with numerous bornhardts and koppies separated by less extensive basins of deeper weathering, whereas the gneisses weather more evenly and give rise to a low, rolling landscape with fewer outcrops. The highest yields are generally from the greenstones (Jordan 1968) while the most consistent results within the granitic rocks are provided by the Younger Granites (Wright 1988).

Field studies were undertaken to help understand the nature of lineaments and their hydrogeological role. However, except in the case of dykes and quartzite veins, which are easily identified at the surface by rock outcrop or scree, direct observation of presumed fracture-related lineaments is seldom possible because the majority are overlain by regolith or superficial deposits. Outcropping bornhardts and koppies provide indirect information but such exposures may represent relatively unfractured kernels atypical of the adjacent rocks beneath the regolith. Nevertheless, this is often the only source of information and the following summary is based largely on the examination of such exposures.

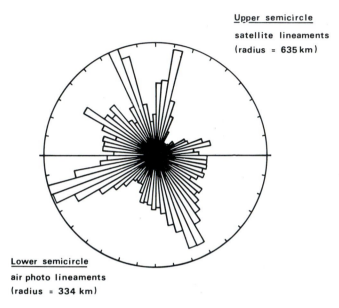

Upper semicircle

satellite lineaments
(radius = 635 km)

Lower semicircle

air photo lineaments
(radius = 334 km)

Fig. 2. Length–frequency rose diagrams for SE Zimbabwe. The figure is a composite derived from an analysis of Landsat imagery (upper semicircle) and 1:80,000 aerial photography (lower semi-circle).

The majority of observed fractures are steep to vertical, closed planar fissures sometimes cemented by quartz. Fault displacements are only occasionally seen but the tectonic origin of many fracture sets can be inferred from their regular development and parallelism with known fault directions, or from rare geomorphic offsets seen on imagery. Uncemented joints often have a youthful appearance and are thought to have formed comparatively recently, perhaps through the release of locked-in palaeostresses (related to earlier tectonic events) during present-day uplift and erosional unloading. That such contemporary processes operate is evidenced in surface-parallel exfoliation sheets seen in granite bornhardts and 'castle koppies', and contemporary fractures formed in quarries and road cuts: such horizontal fissures are thought to relate to expansion of laterally-confined rock masses (Folk & Patton 1982).

Tectonic events in SE Zimbabwe

A reconstruction of igneous–tectonic events in the Zimbabwe Craton from late Archaean through to the Phanerozoic, based on remote sensing, field studies and past mapping (Greenbaum 1987), indicated that the geological development of the Craton was punctuated by repeated episodes of compressive tectonism involving at least four periods of important wrench faulting separated by relaxation and dyke emplacement. This work concluded that most of the lineaments in the Masvingo area were faults or tectonically related joints. Because faults of all origins (normal, reverse and wrench) form as planes of tectonic shearing, such fractures must be regarded as compressive features at least at the time of their inception. As such, they might be expected to act as closed structures from a groundwater viewpoint. Similar considerations have led workers elsewhere to develop 'hydro-tectonic' models which seek to determine the original stress regime on the basis of lineament directions and then use this to identify favourable fracture sites. One such model, for example, tested in Botswana (VIAK 1983 in Buckley & Zeil 1984) attempted to identify tension fractures that sometimes form contemporaneously with the main planes of shearing parallel to the maximum principal stress. Models of this type, however, seem highly suspect. Resurgent tectonism as evidenced in Zimbabwe is likely to have also occurred in other cratonic areas resulting in the reactivation of fractures under new stress conditions and causing a change in their earlier compressive or tensile nature.

Borehole correlation analysis

It is generally assumed that fractures and their associated weathering zones are the principal causes of permeability in basement rocks; consequently, photolineaments are a focus of attention in groundwater exploration surveys. Despite this, little is known regarding the width of fissured zones or the relevance of lineament length or trend to the water-bearing potential of a fracture. In an attempt to study this empirically, an analysis was carried out to compare the yields of existing boreholes with these spatial parameters.

Yield vs distance. Photogeological re-interpretation of almost 300 borehole sites (both producing and dry), in 11 study areas covering a range of geology, found that the majority (more than 90%) were located within 150 metres of a lineament, thus confirming the importance placed on these features in siting. However, an analysis of the drilling results revealed no significant difference in success rates between drill holes located with reference to lineaments and those located without reference to structural information. Thus, although lineaments form the favoured targets, other criteria can provide successful results in areas where no such features occur. For the majority of sites that do fall close to a lineament, attempts to correlate the yield with recorded distance to the lineament were inconclusive, possibly due to inaccuracy in the recorded borehole site locations and/or unreliable pump-test data.

Yield vs azimuth. It is important to know whether any preferred lineament directions exist that provide a lower risk of failure than others. To evaluate this, the correlation between borehole yield and the trend of lineaments occurring within about 150 m was examined. Although the 'zone of influence' around a fracture might be expected to be less than this (Clark 1985), this large margin of error allowed for locational imprecision and ensured that all fracture-related boreholes were included in the analysis. Stacked histograms showing the number of boreholes occurring within each

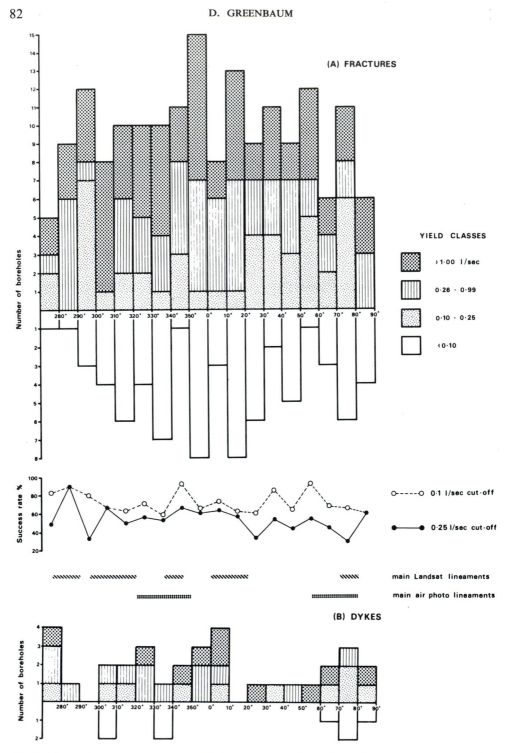

Fig. 3. Stacked frequency histograms for borehole data from eleven detailed study areas in SE Zimbabwe. The plots show the relationships between yield, divided into four sub-classes, and azimuth of associated lineaments. Dominant lineament trends from Fig. 2 are also indicated. Fracture-related lineaments are shown in A and dyke-related lineaments in B.

of four yield classes against lineament directions in 10 degree azimuth classes were constructed for each sub-area and for the total dataset. Fracture- and dyke-related lineaments were considered separately. The limited number of data values make conclusions based on individual sub-area plots unreliable: therefore only plots for the total dataset (Fig. 3A: $n = 248$ fractures; Fig. 3B: $n = 41$ dykes) are reproduced and discussed here.

Figure 3A shows that boreholes have been drilled along fracture-related lineaments of all directions with both successes and failures. The non-uniform distribution of lineament directions needs also to be considered: therefore the dominant fracture trends from Fig. 2 are indicated on Fig. 3. It is apparent from this that only limited correlation exists between successful boreholes and the main lineament directions.

Two measures of success rate were tested based on cut-off flows of 0.1 l/s (this yield class is plotted below the abscissa on Figs 3A & 3B) and 0.25 l/s, which may be considered a more realistic figure for handpump usage. Using this latter cut-off, success rate values better than 60% for fracture-related lineaments (compared with an overall mean of 54%) were obtained for the following directions:

$$280°–290°$$
$$300°–310°$$
$$340°–10°$$
$$80°–90°$$

The equivalent plot for dyke-related lineaments (Fig. 3B) contains fewer data and is therefore more difficult to interpret. Most azimuths are represented and provide reasonable results. In general, few unsuccessful boreholes are related to dykes.

Discussion of results

The correlation studies are interesting but do not convincingly demonstrate the existence of any favourable 'open' fracture direction(s). This confirms the earlier results of Greenbaum (1988) and Lewis (1990). Despite the tendency in the fracture-related dataset for slightly better success rates to occur along approximately west–northwesterly and northerly lineament directions, successful boreholes in fact occur in association with all lineament trends.

In assessing the results, the nature of the data used must be considered. The boreholes were sited and drilled by various agencies over several years and their reliability, both with regard to locational accuracy and pump-test results, is thus uncertain. Although the majority appear to have been sited on the basis of a lineament, the precise relationship is likely to cover a range of situations. Many lineaments are represented at the surface by a drainage line and in such circumstances it is not uncommon for the borehole to be placed some distance away to avoid the possibility of surface flooding: here, a low yield might simply represent a failure to intersect the main zone of fracturing. Our present knowledge does not allow us to reliably interpret the physical characteristics of fracture zones such as dip, width and secondary fissuring from either remote sensing or conventional ground geophysics. These are the subject of ongoing research.

Conclusions

If any general conclusion can be drawn it is that fractures of all orientations are capable of providing permeability and possibly storage, at least in this part of the Zimbabwe Craton. This is an important result but one that requires some additional comment. Despite the geological evidence that most fractures formed as compressive shears, many now appear to provide open conduits for groundwater. The evidence for modern tensional fracturing has already been described. Studies of in situ stress in southern Africa and elsewhere (Price 1966; Van Heerden 1968; Gay 1975, 1977, 1980) support this observation and suggest that, as a result of gravitational unloading, temperature changes and lateral expansion during uplift and erosion, the near-surface zone (down to several tens or possibly hundreds of metres) is one where tensile stresses are dominant. Studies in several countries of Africa and elsewhere suggest that the optimum depth for boreholes in basement ranges from 40 to 80 metres (Davis & Turk 1964; Landers & Turk 1973; Clark 1985), and this is borne out by the results from SE Zimbabwe. In this depth zone, existing fractures (of whatever origin) have a tendency to open and new fractures to form (some representing stored palaeostresses or ancient incipient fractures) including surface-parallel unloading sheets. This means that within the depth zone of most interest to basement hydrogeologists fractures of all types and directions are potentially permeable. The lack of any favoured lineament directions in the present study supports this hypothesis.

Finally, it is interesting to note that seismic studies in southern Africa (Fairhead & Henderson 1977) show that Precambrian crustal structures appear to have an influence on modern crustal movements and suggests that NNE fractures might show increased openness. If this is so, it would seem that such effects are only significant at depths greater than those considered here.

This work was undertaken as part of a research and development project on basement aquifers funded by the Overseas Development Administration. The paper is published by permission of the Director of the British Geological Survey (NERC).

References

BLISS, N. W & STIDOLPH, P. A. 1969. *A Review of the Rhodesian Basement Complex*. Special Publication of the Geological Society of South Africa, **2**, 305–333.

BUCKLEY, D. K. & ZEIL, P. 1984. The character of fractured rock aquifers in eastern Botswana. *In*: WALLING, D. E., FOSTER, S. S. D. & WURZEL, P. (eds), *Challenges in African Hydrogeology and Water Resources*, Proceedings of the Harare Symposium, July 1984, IAHS Publ. **144**, 25–36.

CLARK, L. 1985. Groundwater abstraction from Basement Complex areas of Africa. *Quarterly Journal of Engineering Geology*, **18**, 25–34.

COWARD, M. P. 1976. Archaean deformation patterns in southern Africa. *Philosophical Transactions of the Royal Society of London*, A, **283**, 313–331.

——, JAMES, P. R. & WRIGHT, L. 1976. Northern margin of the Limpopo Mobile Belt, southern Africa. *Bulletin of the Geological Society of America*, **87**, 601–611.

DAVIS, S. & TURK, L. J. 1964. Optimum depth of wells in crystalline rocks. *Groundwater*, **2**, 6–12.

FAIRHEAD, J. D. &. HENDERSON, N. B. 1977. The seismicity of southern Africa and incipient rifting. *Tectonophysics*, **41**, T19–26.

FOLK, R. L. & PATTON, E. B. 1982. Buttressed expansion of granite and development of grus in central Texas. *Zeitschrift für Geomorphologie*, **26**, 17–32.

GAY, N. C. 1975. In-situ stress measurements in southern Africa. *Tectonophysics*, **29**, 447–459.
—— 1977. Principal horizontal stresses in southern Africa. *Pure and Applied Geophysics*, **115**, 3–10.
—— 1980. The state of stress in the plates. *In*: Dynamics of Plate Interiors, Geodynamics Ser., **1**, *American Geophysics Union*, 145–153.
GREENBAUM, D. 1986. *Tectonic Investigation of Masvingo Province, Zimbabwe, Preliminary Report.* BGS Technical Report, MP/86/2/R.
—— 1987. *Lineament Studies in Masvingo Province, Zimbabwe.* BGS Technical Report MP/87/7/R.
—— 1988. *Basement Aquifer Project: Report on Structural Studies 1987/88.* BGS Technical Report WC/88/17.
—— 1989. Remote sensing studies in southeast Zimbabwe. *In: 'Groundwater exploration and development in crystalline basement aquifers'*: Proceedings of workshop held in Harare June 1987, Commonwealth Science Council, *CSC(89)WMR-13*, Technical paper 273, 351–369.
JORDAN, J. N. 1968. Ground-water in the Rhodesian Basement Complex. Geological Society of South Africa annnex to v. LXXXI, 103–111.
LANDERS, R. A. & TURK, L. J. 1973. Occurrence and quality of groundwater in crystalline rocks of the Llano area, Texas. *Groundwater*, **11**, 5–10.
LEWIS, M. A. 1990. The analysis of borehole yields from basement aquifers. *In: Groundwater Exploration and Development in Crystalline Basement Aquifers: Proceedings of workshop held in Harare June 1987*, Commonweath Science Council, *CSC(89)WMR-13*, Technical paper 273, 171–202.
MACGREGOR, A. M. 1947. An Outline of the Geological History of Southern Rhodesia. *Bulletin of the Geological Society of Southern Rhodesia*, **38**.
—— 1951. Some milestones in the Precambrian of Southern Rhodesia. *Proceedings of the Geological Society of South Africa*, **54**, 27–71.
PRICE, N. J. 1966. *Fault and Joint Development.* Pergamon, Oxford.
STOWE, C. W. 1980. Wrench tectonics in the Archaean Rhodesian craton. *Transactions of the Geological Society of South Africa*, **83**, 193–205.
VAN HEERDEN, W. L. 1968. *Measurement of Rock Stresses at Shabani Mine, Rhodesia using a C.S.I.R. Triaxial Strain Cell.* South African Council for Scientific and Industrial Research, Pretoria, Rep. MEG 663.
WILSON, J. F. 1973. Granites and gneisses of the area around Mashaba. *Geological Society of South Africa, Special Publication*, **3**, 79–84.
—— A preliminary reappraisal of the Rhodesian Basement Complex. *In*: ANHAEUSSER, C. R., FOSTER, R. P. & STRATTEN, T. (eds) A symposium on mineral deposits and the transportation and deposition of metals. *Geological Society of South Africa, Special Publication*, **5**, 1–23.
——, JONES, D. L. & KRAMERS, J. D. 1987. Mafic dyke swarms in Zimbabwe. *In*: HALLS, H. C. & FAHRIG, W. F. (eds) *Mafic Dyke Swarms*, Geological Association of Canada, Special Paper, **34**, 433–444.
WRIGHT, E. P. 1988. *Basement aquifer project. Project report for 1987–1988.* BGS Technical Report.

From WRIGHT, E. P. & BURGESS, W. G. (eds), 1992, *Hydrogeology of Crystalline Basement Aquifers in Africa*
Geological Society Special Publication No 66, pp 87–100.

An exploration strategy for higher-yield boreholes in the West African crystalline basement

Eduard Boeckh

Birkenweg 10, Wettmar, D-3006 Burgwedel 5, Germany

Abstract. In Burkina Faso investigations have been carried out to test satellite imagery and electromagnetic (EM) methods as a means of improving the siting of water supply boreholes in fissured rocks of the crystalline basement. The results show that these methods may be used to locate boreholes with above average groundwater yields, and to reduce the failure rates.

Geometrical analysis of Landsat–MSS lineaments reveals the relevant regional fracture pattern, the present-day principal stress field and the resulting system of oriented hydraulic conductivities. Structural analysis of the lineament pattern and identification of dry-season-active vegetation on Landsat-TM imagery permit promising locations to be pinpointed for groundwater abstraction. EM-measurements on a two-dimensional grid using the 'very low frequency' (VLF) and 'horizontal loop EM' (HLEM) methods are used to locate potential fractures on a local scale as a basis for rational borehole siting.

The methods, applied separately or in combination, offer an economical approach for use in rural and community water supply, small-scale irrigation and water resources planning. The methods appear to be generally transferable to fissured hard rock aquifers in other regions. The best suited settings are believed to be on cratons and platforms in a semi-arid climate.

Every year thousands of boreholes are drilled into the crystalline basement of West Africa (Fig. 1) with the intention of constructing wells for rural and community water supplies. Because the existing shallow porous aquifers, regolith and local alluvial deposits normally have a low yield, most of the boreholes are sunk into the hard rock in order to tap groundwater in fractures. Many boreholes, however, are unsuccessful.

The most common reason for such failures is that the unsuccessful boreholes do not encounter water-bearing permeable fractures. In central Burkina Faso for example, only four boreholes out of five produce enough water to meet the requirements of a rural water supply, i.e. a minimum of about $0.5 \, m^3/h$. For the purpose of a centralized community water supply on the other hand, where a well equipped with a motor pump must have a minimum yield of about $5 \, m^3/h$, several boreholes usually have to be drilled before one produces the quantity of water required. Considering these facts one may ask: can the success-rate be enhanced by improving the technique of siting boreholes?

The method currently used for prospecting for groundwater and siting boreholes in the crystalline basement of West Africa employs air photographs and field inspection to recognize zones of possible rock fracturing. In favourable regions this procedure is occasionally supplemented by geoelectrical measurements. The fractures located in this way are not always water bearing. The ultimate result of drilling, in terms of groundwater yield, is uncertain.

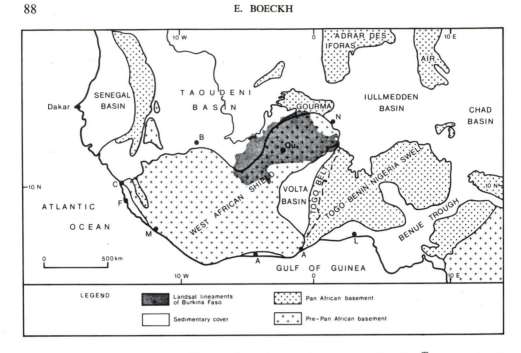

Fig. 1. Geological sketch map of West Africa showing the outlines of the West African shield and the study area within the confines of Burkina Faso where lineaments of Landsat Multi Spectral Scanner (MSS) imagery provide the basis for structural analysis.

Additional information concerning structure and fracturing may now be provided by satellite imagery and electromagnetic geophysical methods. These methods provide a better definition of the fractures before they are explored by drilling.

(i) Imagery produced by the satellite system Landsat Multi Spectral Scanner (MSS) can, under certain conditions, display structural information which can be interpreted in terms of present-day elastic deformation and hydraulic conductivity.

(ii) Satellite imagery at the same time provides information on the vegetation, which under certain conditions may indicate groundwater occurrences including productive fractures. Valuable information on an appropriate scale for this purpose is also provided by the satellite system Landsat Thematic Mapper (TM). Other advantages of this system are mentioned later.

(iii) Electromagnetic measurements permit a rapid survey of a favourable region to be made revealing details of the local fracture pattern.

The value and applicability of these methods has been tested in an investigation carried out in a typical area of the crystalline basement in Burkina Faso.

Scope of the research

The specific aims of the investigation were:

(i) to test the suitability of the methods noted above in regions directly underlain by crystalline basement;

(ii) to adapt the methods to the specific environment of Burkina Faso, estimating their individual advantages and defining their requirements;

(iii) to determine the type of area in which the methods are most applicable, and their transferability to other regions.

The area chosen for the investigation is situated in central Burkina Faso north of the capital Ouagadougou. In this area the crystalline basement is made up of gneisses, granites, migmatites and schists and is part of a single tectonic block, the West African Shield (Fig. 1). The surface conditions are suitable for reconnaissance of structural features by means of remote sensing. An interpretation of lineaments mapped on the basis of Landsat MSS was available (Fig. 2). Project reports covering many existing boreholes and wells were available outlining the general hydrogeological conditions. Air photographs, topographical maps and Landsat TM imagery were also available.

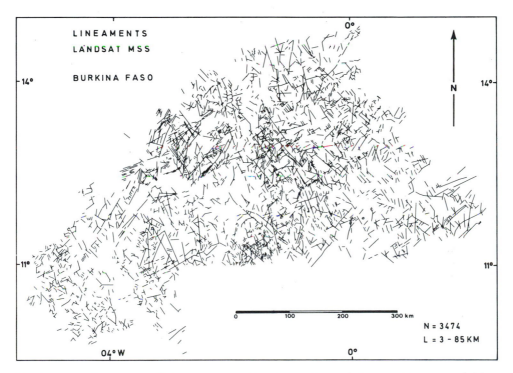

Fig. 2. Map of Landsat MSS lineaments within the confines of Burkina Faso recorded by Bannert *et al.* (1980). The lineaments have been visually mapped on black-and-white Landsat MSS imagery on the scales of 1:500 000 and 1:1 000 000. Most of the area of 274 000 km^2 is made up of crystalline basement belonging to the West African shield (Fig. 1).

In the selected area the studies were concentrated on four representative communities. In each case the preparatory work involved remote-sensing surveys at a variety of scales as well as field studies covering geological, hydrogeological and geophysical surveys. The field studies were supported by the drilling of 16 boreholes to a final depth of 80 m and air-lift and pumping tests.

The results of the studies are presented below by considering three principal methodological aspects.

Results of the investigations

Hydraulic conductivity

The proposed method of determining the kind of tectonic deformation and resulting orientation of the hydraulic conductivity of the rock mass is based on the map of the Landsat MSS lineaments (Fig. 2), which owing to the low resolution of the satellite recording system represents major lineaments only. The length of the lineaments ranges from 3 to 85 km. The density of the lineaments appears to depend on lithology and regolith. There are more than 3000 lineaments and many intersect each other.

The method applied comprises geometrical analysis of intersecting lineaments. It is assumed that the lineaments represent tectonic structures of the same age and origin and that it is possible to deduce the orientation of the stresses responsible (Ragan 1985) and also to draw conclusions regarding the orientation of the hydraulic conductivity in the rock mass. The approach is empirical.

An analysis is made of pairs of lineaments plainly intersecting each other (x-type intersections). Altogether about 650 x-type intersections were identified. The parameters considered are the angles of intersection and directions of the intersecting lineaments. For simplicity the directions of the two intersecting lineaments are replaced by a single direction of the bisectrix.

The results of the analysis are as follows. The diagrams of angle of intersection versus direction of bisectrix (Fig. 3) reveal properties characteristic of a statistical population of conjugate shears formed by a distinct and uniform tectonic event. The clusters in the scatter diagram (Fig. 3a) correspond to the four sectors of a set of two intersecting lineaments. The isoline diagram (Fig. 3b) shows a concentric pattern around a distinct maximum; it is symmetrical with respect to direction. The contours occupy different ranges of conjugate lineament angle in neighbouring quadrants.

It can be shown from the concept of conjugate shears (Ragan 1985) that the axis of symmetry in the quadrants of the dominant acute angles of Fig. 3 corresponds to the axis of greatest principal stress σ_1. The axis of symmetry in the quadrants of the dominant obtuse angles perpendicular to σ_1 on the other hand corresponds to the least principal stress σ_3. The respective directions in this case are

$$\sigma_1 = N\ 005°$$
$$\sigma_3 = N\ 095°.$$

A histogram of the angles of intersection in one of the quadrants of σ_1 (Fig. 4) reveals that brittle as well as ductile deformation has affected the rock mass. The histogram shows two peaks at angles of 90° and near 60°; these angles are typical for conjugate shears formed under ductile and brittle conditions respectively (Jaroszewski 1984). The structures originating from ductile deformation are believed to have been formed at deeper levels of the crust and to have been transferred mechanically to the surface.

The ductile conjugate shears appear to have the character of strike-slip shears which would suggest that they are vertical with respect to the earth's surface. This follows from the fact that the angles represented by the maxima in the four quadrants of Fig. 3b are the same, i.e. 90°, as would be expected if two vertical planes intersect each other perpendicularly. The brittle conjugate shears are also likely to be vertical or near vertical.

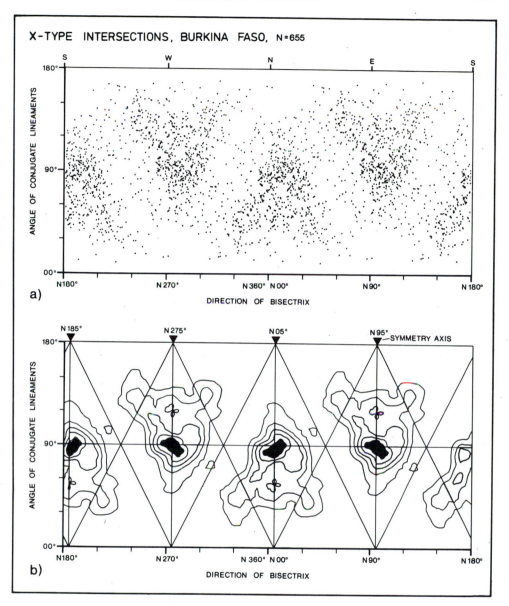

Fig. 3. Direction of bisectrix versus angle between conjugate lineaments for x-type intersections seen on Landsat MSS images, Burkina Faso; (a) scatter diagram; (b) isoline diagram. The symmetrical pattern is the result of a distinct tectonic deformation. It permits the configuration of the principal stress field to be determined.

Brittle deformation is characterized by a spread in the trends of the conjugate shears. Ruhland (1973) has suggested that brittle deformation causes derivative shears of several orders to form successively more deviating trends (Fig. 5). According to his scheme the spread of lineament trends in the brittle range of deformation in the direction-versus-angle diagram (Fig. 3) appears to be produced by the first two orders of conjugate shears. In this diagram the conjugate shears form individual submaxima or families (Fig. 6).

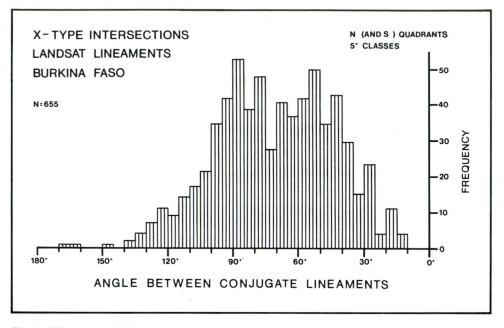

Fig. 4. Histogram of the angles of the x-type intersections between conjugate lineaments in one of the σ_1-quadrants in Fig. 3 characterized by dominant acute angles. The double peak reflects two subpopulations, one of which is formed by ductile and the other by brittle deformation.

Table 1. *Families of conjugate shears recorded on Landsat MSS imagery of Burkina Faso*

Number of family*	Mode of deformation	Order of deformation	Sense of displacement on shears†		Angle	Direction of σ_1
1	ductile	I	Opposite senses: sinistral and dextral		85° (±20°)	N 005° (±20°)
2					55° (±15°)	N 005° (±15°)
3					74° (±10°)	N 022° (±10°)
4	brittle	II			62° (±10°)	N 350° (±10°)
5			same senses:	dextral	42° (±20°)	—
6				sinistral	40° (±15°)	—

* refers to the families in Fig. 6.
† of x-type intersections.

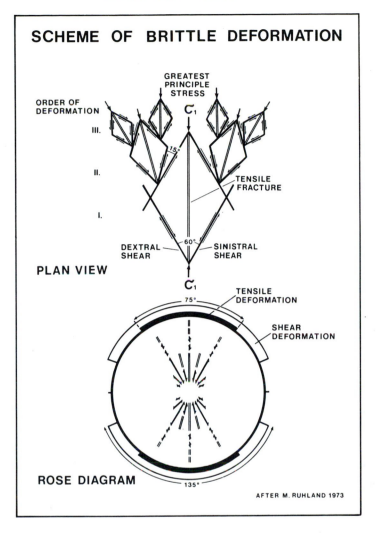

Fig. 5. Scheme of brittle deformation of a homogeneous rock mass, modified after Ruhland (1973). The deformation produces fractures of different magnitude and direction in successive orders. The rose diagram illustrates the orientation of the various fractures distinguished by character and magnitude.

Table 1 presents the mode of formation and geometrical data for the various families of conjugate shears identified in Fig. 6. The deviations in direction and angle are due to anisotropy of the rock mass attributed to lithological variations and pre-existing fractures.

The process of shearing is accompanied by tensile deformation. Tensile fractures according to the scheme are expected to form in the direction of the greatest principal stress σ_1 associated with each order of brittle deformation (Fig. 5). It is proposed that the nature of these tensile fractures is that they tend to increase the hydraulic conductivity.

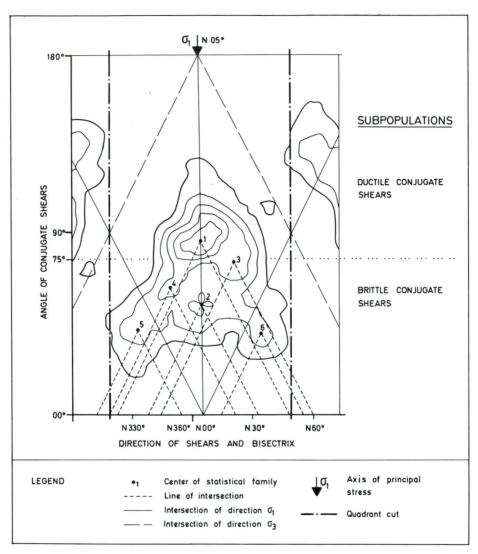

Fig. 6. Subpopulations and families of x-type intersections of Landsat MSS lineaments in Burkina Faso in one of the σ_1-quadrants of the direction-versus-angle diagram (Fig. 3). The subpopulations are characterized by distinct geometrical features. The families (1–6) appear to be formed by conjugate shears belonging to two successive orders (Table 1).

In a rock mass fractured by three orders of brittle deformation, a situation likely to be realized on a local scale, tensile fracturing and the resulting increased hydraulic conductivity can be expected to spread over a sector of around 75° (Fig. 5). The largest tensile fractures and the corresponding highest hydraulic conductivities are expected to occur parallel to the direction N 005° ($\pm$ approx. 15°). Three orders of shear fractures on the other hand occupy a sector of around 135° by this scheme. The orientation of vegetation lineaments as recorded on Landsat MSS imagery of central Bukina Faso at a scale of 1:200 000 has been observed to show a close correlation with this concept of rock fracturing and related hydraulic conductivity.

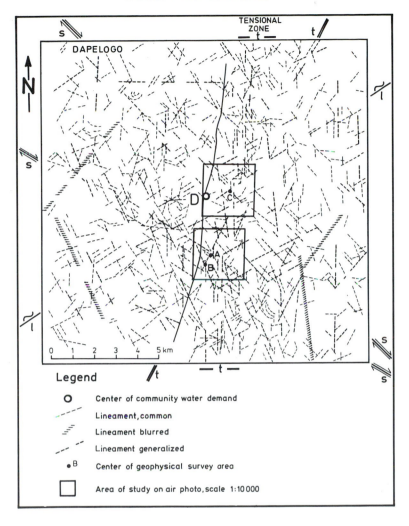

Fig. 7. Map of Landsat TM lineaments in the Dapelogo area, central Burkina Faso; example of reconnaissance for promising locations on the scale of 1:50 000; part of Landsat TM image 195/051 of 3 February 1986 obtained as 'special acquisition' at the time of maximum contrast between active and inactive vegetation in the dry season; vegetation enhanced by false colours; structures visually determined; important structures generalized.

The areas at the intersection of a major N–S tensional zone (t) with important zones of shearing (s) near the centre of demand (D) were chosen for subsequent detailed study on enlarged air photos. The intersections are characterized by very active vegetation in the dry season. The regional strike (1) of the rock formations runs roughly NE–SW. The sites at which geophysical measurements (A, B, C) were made were selected at intersections of important fracture zones on the air photos.

Promising locations

Landsat TM imagery has been tested as a means of searching for productive fracture zones. The specific qualities of the satellite system Landsat TM in this respect are: the structures can be seen in more detail than by Landsat MSS, owing to the higher resolution of the Landsat TM system (Joannes *et al.* 1986) (air photographs on the other hand tend to show too much detail); the structures are undistorted, an essential feature for tectonic interpretations; information on the vegetation which might be related to the presence of groundwater can be enhanced; and images can be reproduced at any scale, e.g. at the scale of existing air photos, and for any part of the area desired.

The area investigated in central Burkina Faso has several features making it suitable for the study: the undulating, partly hilly country is mostly covered with shrubs; trees occur mainly on the plains and in valleys. Land cultivation is seldom intensive. The crystalline basement, mainly composed of migmatites and schists, is fairly uniform; weathering extends to a depth of 10 to 30 m, only locally deeper. Outcrops are scarce. Two of the community sites investigated are situated in a valley, and two in an elevated position near a water divide. Judging from the existing wells and boreholes, the hydrogeological conditions are rather complex.

The procedure of preparing and evaluating the Landsat TM images is demonstrated in connection with the map of Dapelogo, one of the four sites in central Burkina Faso investigated (Fig. 7). The interpretation of the lineament pattern is based upon the concept of regional tectonic deformation explained in the preceding section. In this way the most favourable spots are selected for continued investigation using geophysical measurements and drilling.

The general results of investigation of the four sites based on geophysical measurements and boreholes are as follows: groundwater prospective zones can be readily identified on the basis of structure, geomorphology and vegetation; the most favourable structural conditions are believed to exist at the intersections of major zones of shear and tensile fracturing, which are marked by active vegetation during the dry season; geomorphology has to be taken into consideration to ensure that satisfactory conditions exist for adequate groundwater recharge.

The role of dry season active vegetation as an indicator of groundwater-bearing fractures, however, is not always conclusive. Moisture or groundwater contained in the regolith or local alluvial aquifers independent of major productive fracture zones can also be responsible for stable active vegetation at times. On the other hand, vegetation dependent on groundwater may have been removed where land has been cultivated or degraded. An additional examination of air photos and a field check are therefore indispensable. Another factor that may limit the existence of continuously active vegetation is the depth to the water table, when it is more than about 10 to 15 m below ground surface.

Landsat TM should not be used for borehole siting because of the relatively low resolution of the imagery. Air photographs and geophysical measurements are more appropriate for the purpose.

Fracture network

The third step in the investigation is to test the capability of the EM-methods to reveal the fracture network on a local scale. The results should then serve as the basis for optimum borehole siting.

The EM methods applied are very low frequency (VLF) and horizontal loop (HLEM). They are particularly sensitive to vertical and subvertical fractures which can act as groundwater conduits, a highly important attribute. The EM-methods may be supplemented by magnetic measurements and geoelectrical soundings.

The measurements in each case cover an area of 500×500 m on a close grid of mutually perpendicular lines. The lines are 50 m apart and the measurements are taken every 25 m. The resulting anomalies obtained by the different EM-methods are integrated and interpreted in the light of the preceding structural studies. The drilling sites are then selected and priorities assigned on the basis of the inferred nature of the rock fractures and hydraulic conductivities, potential groundwater recharge, access for the drilling rig, and other relevant criteria. One of the total of ten geophysical survey areas established for the four communities considered is shown as an example in Fig. 8.

The centre of the geophysical survey area has to be fixed in advance of the measurements. This can best be done in relation to the structures recognized on enlarged air photos, scale 1:10 000. The orientation of the survey lines should best be arranged so that one set is parallel to the direction of the highest hydraulic conductivity of the region.

The advantages of EM-methods for borehole siting are as follows:

(i) the two-dimensional record of EM-anomalies provides an inventory of potential fracture zones from which a more rational choice of drilling site can be made;

(ii) their interpretation in the light of the structural analysis of the region permits the most promising borehole sites to be selected.

It is found that the principal structures located are genetically compatible with those recognized by remote sensing. The hydraulic properties which were determined in the basement after the 16 project boreholes had been drilled are found to agree with the results of the structural analysis of the region carried out previously: elevated borehole yields are associated with N–S oriented fractures; high-conductivity fracture zones extend up to about 10 km in a N–S direction.

Not all EM-anomalies, however, are related to zones of rock fracturing; some EM-anomalies coincide with magnetic anomalies and/or geological structures, indicating a lithological origin (Fig. 8).

Furthermore, not all boreholes sited on prospective structures are found to be productive: minor N–S fractures may be due to shearing or to tensile deformation (Fig. 5) and will be hydraulically favourable only in the latter case; minerals may be deposited in open fractures and can thus reduce their hydraulic conductivity; lastly, variability in the hydraulic conductivity within a fracture zone may even cause an apparently well sited borehole to be unsuccessful.

In spite of these uncertainties, application of two-dimensional EM-measurements in connection with structural analysis of the region should in general significantly improve the overall success rate of water supply boreholes. The average yield of boreholes should be greater and the failure rate less.

Applicability of the methods

The methods described are applicable to all the groundwater related problems of the crystalline basement in the region investigated. Their application is expected to cover the fields of rural water supply, community water supply, small-scale irrigation and water resources planning. The methods can be applied separately or in combination.

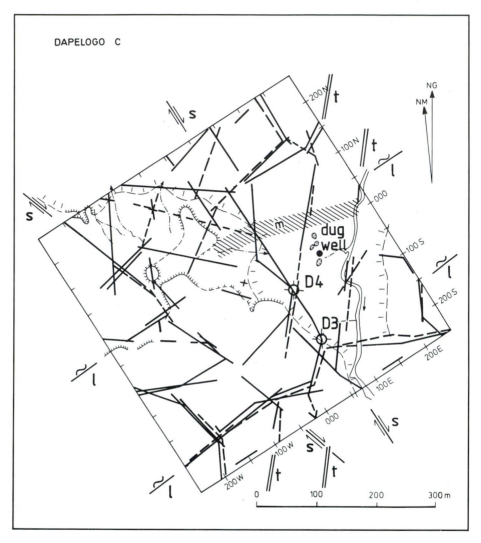

Fig. 8. Map showing the axes of electromagnetic anomalies (thick continuous and dashed lines) and sites of the successful boreholes D3 and D4 in the geophysical survey area of Dapelogo C (Fig. 7); integrated results of VLF and HLEM measurements; example of a typical network of potential fractures in the crystalline basement of central Burkina Faso and related borehole siting. The sites for boreholes D3 and D4 were chosen at points of intersection of important zones of shearing (s) and tensional fracturing (t). Magnetic anomalies (m) coincide with structures of lithological origin (1). Borehole D4 has a regionally outstanding yield of $7\,\mathrm{m}^3/\mathrm{h}$. The flood plain near the boreholes is a favourable area for groundwater recharge.

Rural water supply

The application of the regional structural analysis described above to the interpretation of air photos can in itself improve the success of borehole siting. The average yield and success rate of the boreholes so sited may be greater and the vulnerability of the wells in periods of drought be less. This is valid throughout the territory of Burkina Faso and surrounding crystalline areas.

The methods employing Landsat TM imagery and EM-measurements are too costly to be applied routinely in the field of rural water supply. Their application, however, may be helpful in exceptional cases.

Community water supply

The search for promising sites for high-yield boreholes required by centralized community water supplies will profit most from application of the methods applied in combination. Systematic and successive use of the methods is most effective in focusing the search for the most favourable sites. The elevated cost of Landsat imagery and EM-measurements should be justified by the enhanced borehole yield obtained as well as by savings in drilling.

In cases where the hydrogeological situation is not so complicated, the methods can be adapted as required and can thus substantially reduce the overall cost of exploration.

Small-scale irrigation and water-resources planning

Application of Landsat TM imagery combined with regional structural analysis as described above can contribute substantially to an inventory of potential ground-water resources and, as we have seen, can help with the identification of zones that are favourable for groundwater abstraction. This is also valid for the territory of Burkina Faso and neighbouring areas.

EM-measurements, if used in the later stages of groundwater exploration and adapted to the specific conditions, may be useful in siting boreholes for small-scale irrigation.

Transferability to other regions

Whether the methods described can be transferred to other regions depends essentially on structural, climatic, and lithological factors.

Structure

The types of tectonic deformation recognized in Burkina Faso by the analysis of Landsat lineaments have also been found to exist in central Niger and on the Arabian platform. It is likely that they have affected other cratonic regions and platforms in a similar manner. There is little doubt that geometrical analysis of Landsat lineaments in many such areas could profitably contribute to the understanding of the structures and consequently to the search for groundwater.

Climatic condition

In Burkina Faso the surface conditions are very favourable for carrying out a structural analysis based on Landsat lineaments: the vegetation is not so dense that the structures are obscured. To some extent the existing sparse vegetation appears to enhance the lineaments. Vegetation that remains active throughout the dry season can act as a useful indicator of the presence of groundwater.

Experience suggests that structural studies of the kind described above are likely to succeed under similar surface conditions elsewhere. Conditions of this sort exist in wide semi-arid belts in Africa and other continents.

Lithology

The pattern of Landsat lineaments depends to some extent on the rock type. The lineaments characteristic of the sandstone formations of southern Mali, for example, are generally larger, more persistent and more widely spaced than those recorded on the crystalline basement in neighbouring Burkina Faso, although both areas belong to the same cratonic unit. Structural investigations of this kind therefore have to be adapted to the dominant rock type in the area.

However, in some cases it may be found that a rock type, because of its electrical properties, is not amenable to the EM-methods. Electrical conductivity in the sandstone formation of southern Mali, for example, is too small for anomalies to show up, even in the case of fracture zones which have been detected by remote sensing and proved by drilling.

The project was made possible and funded by the German Ministry of Economic Cooperation. Sincere thanks are due to many of my colleagues at the Bundesanstalt für Geowissenschaften und Rohstoffe, Hannover for contributions and assistance. The cooperation of the Comité Interafricain d'Etudes Hydrauliques (CIEH), Ouagadougou is gratefully acknowledged.

References

BANNERT, D., HOFFMANN, R., JOENS, H. P. et al. 1980. Etude de reconnaissance des ressources en Afrique à l'aide d'images de satellite en République de Haute-Volta: Géologie, Hydrogéologie, Pédologie et Utilisation du sol. Bundesanstalt für Geowissenschaften und Rohstoffe, Hannover.

JAROSZEWSKI, W. 1984. *Fault and Fold Tectonics*. Ellis Horwood, Chichester.

JOANNES, H., PARNOT, J., RANTRUA, F. & SOW, N. A. 1986. Possibilité d'utilisation de la télédetection dans le domaine de l'eau en Afrique. *Série Hydrologie Comité Interafricain d'Etudes Hydrauliques*. Ouagadougou.

RAGAN, D. M. 1985. *Structural Geology* 3rd edn, Wiley, New York.

RUHLAND, M. 1973. Méthode d'étude de la fracturation naturelle des roches associée à divers modèles structureaux. *Bulletin des Sciences Géologiques*, **26**, 91–113.

From WRIGHT, E. P. & BURGESS, W. G. (eds), 1992, *Hydrogeology of Crystalline Basement Aquifers in Africa*
Geological Society Special Publication No 66, pp 101–129.

Groundwater movement and water chemistry associated with weathering profiles of the African surface in parts of Malawi

M. J. McFarlane

School of Geography, Oxford University, Mansfield Road, Oxford OX1 3TB, UK

Abstract. This paper summarizes the main results of recent geomorphological research on the hydrogeology of weathering profiles on the African erosion surface in Malawi. Deep regolithic profiles have developed by protracted and aggressive weathering and differential leaching. In the advanced stages of weathering, congruent kaolinite 'dissolution' causes the saprolite to collapse, forming a thick residuum of the most resistant materials, dominated by silica in the form of quartz and iron as goethite. Aluminium has been extensively leached. Such profiles pertain to a mechanism of land surface formation dominated by the activity of infiltrating water rather than direct surface runoff and it results in terrain with a generally basined configuration. Low-lying areas are occupied by dambos. These are clay-filled bottomlands which are seasonally waterlogged and are located where lithology favours leaching and saprolite collapse. The contemporary fluviatile-like configuration of the dambos is attributed to post-incision modification of the ancient land surface. As a result, their area is reduced and they become inset into it, following geological controls which express integrated groundwater movement within the saprolite. The infill is shown to be essentially an alumino-silicate evaporite rather than of alluvial origin.

Analysis of dambo and dambo-peripheral profiles supports this genetic model, identifying the continuity of the contemporary dambo infill with a 'palaeodambo' clay wedge below the surficial sands in peripheral situations. The very low permeability of the clay is held to be responsible for the marginal seepage zone, fed by shallow throughflow from topographically higher profiles and augmented by upward discharge of deep water.

Elements lost from the interfluves were identified in evaporites which occur on the dambo floors in the dry season. These include aluminium. Analysis of water discharging into the dambos failed to identify this element. There is a need for reassessment of techniques to determine element mobilization in tropical groundwater, where organic binding is implicated.

Background

The African erosion surface in Malawi is characterized by low relief, with areas of seasonally waterlogged bottomlands known as dambos. This ancient landsurface has been exposed to aggressive and protracted weathering and leaching which has resulted in a thick mantle of regolith. The weathering profiles have very high primary porosity. The resulting high storativity is extremely important to groundwater supplies in areas of crystalline basement rocks with negligible primary porosity. Recognition of this prompted attempts to understand better the nature of the circulation systems within the regolithic mantle. Information on the variability of the weathering profile horizons in different catenary positions is essential to this because in such low relief contexts with relatively small heads on the circulation systems quite subtle variations, for example in clay content, can be expected to exert a strong influence on water flow pathways.

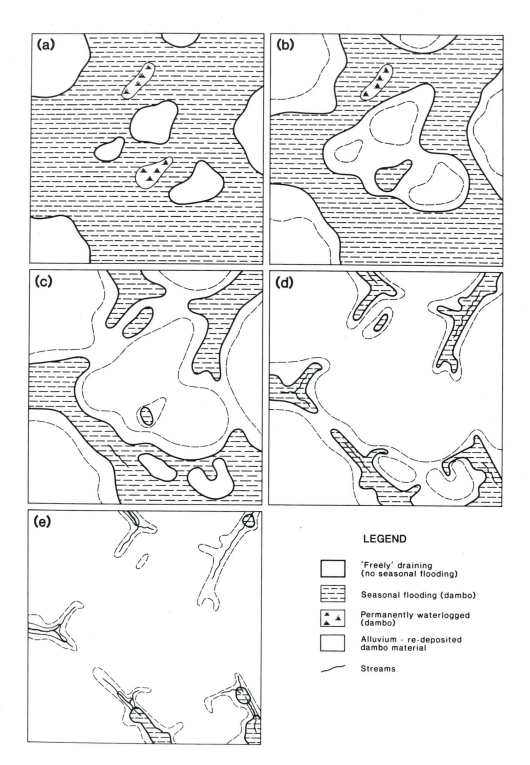

LEGEND

☐ 'Freely' draining
(no seasonal flooding)

▭ Seasonal flooding (dambo)

▣ Permanently waterlogged
(dambo)

▢ Alluvium - re-deposited
dambo material

⌒ Streams

The African erosion surface

There are two fundamentally different hypotheses concerning the origin of the surface. The literature on Malawi (e.g. Bellingham & Bromley 1973) follows King (1963, 1967) and Lister (1967) in ascribing the low relief to a mechanism of 'stripping' associated with the process of pediplanation. This effectively attributes its erosion to the activity of direct surface runoff. The alternative process, which may be described by the ill-defined term etchplanation (Wayland 1934), involves differential leaching (Büdel 1957, 1977; Aleva 1983). As a result, weathered rock (saprolite) collapses to form a thick surficial residuum or colluvial mantle of materials which is relatively resistant to leaching, a blanket which may be locally redistributed to even out the 'highs' and 'lows' of the irregularly lowered surface. In general terms, relief becomes subdued by a process of downwasting which involves the slow subsidence and migration of residual materials downslope towards the bottomlands. In this case the erosion surface is made essentially by the activity of infiltrating water. These processes, pediplanation and etchplanation, are not exclusive. Where the climate is dry and vegetation cover poor, direct surface runoff is more able to 'strip' residua and, since the ancient cratonic areas have suffered both climatic change and variable migration with respect to global climatic belts, their surfaces may well express the effects of both processes.

The dambos

The origin of the dambos is also a matter of some debate. In broad terms, two conflicting models can be identified. In the fluviatile model they are viewed as ancient river systems, their valleys excavated by direct surface runoff and subsequently infilled under conditions of progressively reduced flow energy, to give an infill which is a 'fining-upwards' succession, with gravel and sand at the base and clay at the top (Freshney 1985, 1987). An alternative model has been proposed, in which the dambos are the bottomlands of the landsurface, which was irregularly lowered by differential leaching (McFarlane 1989). The development of their predominantly integrated patterns, as seen today, is considered to be a stage in the destruction of the dambo system, contingent upon uplift and incision of the ancient surface; the fluviatile-like configuration is attributed to integrated groundwater flow beneath the clayey infill. This model of dambo evolution is summarized in Fig. 1. The progressively shrinking area of a dambo would leave a 'palaeodambo' clay wedge at the

---

Fig. 1. Tentative model of the evolution of dambo forms associated with incision and water table lowering. (a) Original deeply weathered landsurface of very low relief. Waterlogged lowlands have a reticulated pattern, surrounding freely-draining land. (b) Incision lowers the regional water table, reducing the area of waterlogged lowland. (c) Further water table lowering breaks the dambo links across divides, initiating an integrated pattern. (d) With ongoing incision and water table lowering, the narrowing dambos become linear, expressing geological control. (e) The integrated dambo system breaks up into small stretches of dambo (more deeply weathered), separated by valleys with streams and outcropping rock. Reworked dambo infill forms an alluvial strip flanking the streams. Ultimately, stream incision completely destroys the dambos.

margins of the contemporary dambo and this evokes a quite different hydrogeological scenario. For example, in the case of the fluviatile model, it might be possible
for direct surface runoff or shallow throughflow from the interfluves to recharge the
groundwater system via the sandy base of the dambo infill (Fig. 2). In the second
model, the clay wedge would prevent this.

Thus the models of the origins of the erosion surface and dambos are genetically
linked; pediplanation and a fluviatile dambo origin imply distinctly different hydrological and hydrogeological processes from those implicit in the etchplanation model
which attributes both the origin of the surface and the origin and evolution of the
dambos to ingress of surface water and its function as a subsurface agent. In this
study therefore the processes of surface and dambo formation are central issues,
contributing to the understanding of water movement systems and associated
variations in water chemistry.

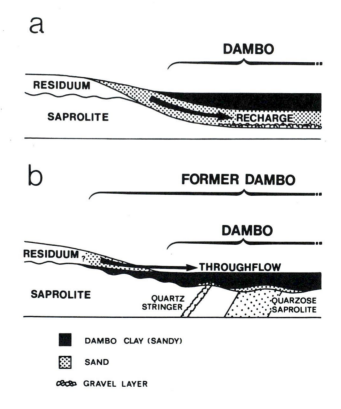

Fig. 2. Hypothetical stratigraphical relationships between dambo 'sediments' and the regolith
profile. (a) Fluviatile erosion truncates the regolithic profiles, comprising residuum or
colluvium overlying saprolite. Fluviatile sedimentation under progressively reducing energy
conditions results in a 'fining upwards' succession. Stratigraphic continuity of the marginal
sands with the sub-clay sands provides the opportunity for direct surface runoff or shallow
throughflow to recharge deeper groundwater beneath the clay. (b) Incision of a low relief
surface, with clay-filled bottomlands, and reduction in dambo area leaves a 'palaeodambo'
clay wedge flanking the contemporary dambos. Throughflow in the sand (a residuum from the
leaching of the exposed infill) discharges at the margins of the dambo floors.

Location of the study area

In order to examine these genetic issues, profiles associated with two catenas, from interfluves into the dambo bottomlands, were examined. The Linthembwe dambo catena, near the village of Chikhobwe in Dowa District includes a relatively narrow dambo, deeply inset into the surrounding terrain (Fig. 3). By contrast the Chimimbe dambo catena near the village of Magomero in Kampini District includes a much more extensive dambo only slightly inset into the erosion surface (Fig. 4).

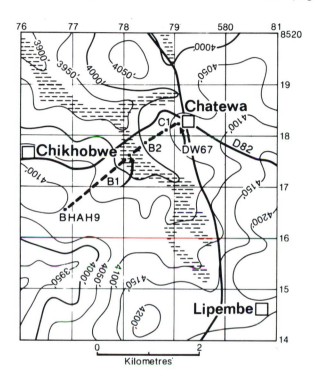

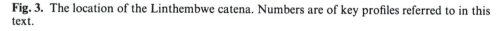

Fig. 3. The location of the Linthembwe catena. Numbers are of key profiles referred to in this text.

Profiles

Interfluve profiles

The interfluves have a generally reddish, clayey sand surface mantle. Borehole drilling logs commonly record a change of colour to grey or greenish grey saprolite at depths of 10–20 m, first water strike often co-inciding with this change. Rest levels lie some metres above this, even in interfluve situations, indicating a semi-confined system with lower permeability in the reddish coloured upper parts of the saprolite.

Chimimbe interfluve profile (B3). At this location (Fig. 4) the regolith is some 20 m thick (Fig. 5). The base of the upper component, the earthy reddish sandy colluvium

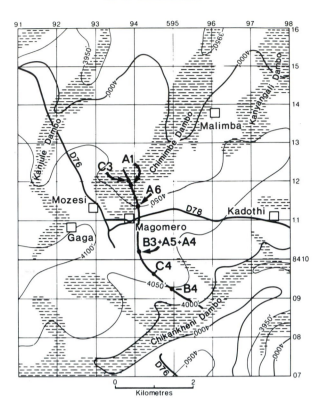

Fig. 4. The location of the Chimimbe catena. Numbers are of key profiles referred to in this text.

or residuum which does not retain original rock textures or structures, is ill defined. Clasts of saprolite increase with frequency from about 5 m downwards until at about 7 m the material is entirely saprolitic. Scattered Fe and Mn pisoliths in the clayey lower part of the residuum indicate a perched, oscillating, water table. The saprolite is divisible into two parts. The upper, variegated, pink, white and yellow coloured material is the zone of 'superior alteration' (de Lapparent 1941; Lelong & Millot 1966), characterized by kaolinization and oxidation of the iron minerals. Their development is associated with vadose conditions. Below this lies green, grey and white coloured saprolite, the zone of 'inferior alteration', associated with phreatic conditions. Mn and Fe mottles and fracture face deposits occur particularly in the lower 3 m of the zone of 'superior alteration', again indicating a zone of water table oscillation. The brecciated zone at the base of the saprolite is deduced, on the basis of the poor core recovery, to be some 3–5 m thick; the material is neither sufficiently clayey to stick together, nor has it sufficient primary mineral bonding to facilitate recovery. The parent rock is a quartzo-feldspathic gneiss, in places amphibolitic and with minor biotite-rich facies.

X-ray diffraction showed (Fig. 5) that the zone of superior alteration and the residuum are dominated by kaolinite and quartz, with Fe in the form of goethite. Interestingly, kaolinization extends well into the zone of 'inferior' weathering where

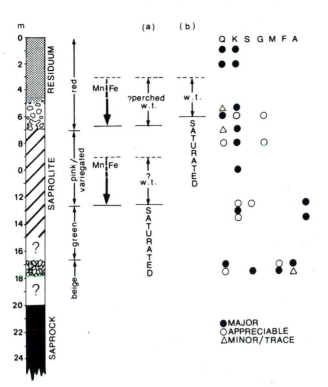

Fig. 5. Main features of the Chimimbe interfluve profile (B3). (a) Former water tables; (b) present water table; Q, quartz; K, kaolinite; G, goethite; S, smectite; F, feldspar; M, mica; A, amphibole.

smectite is also recognizable and there is survival of amphiboles. Mica and feldspar were only identified near the base of the saprolite.

Although these profile characteristics show, with clarity, the hydrogeological conditions under which this profile developed, today the main water table lies nearer the surface, at 3–6 m depth (Allen 1987, 1988). Thus, some 6 m of formerly vadose saprolite has been drowned by a subsequent rise in the water table, this despite the general scenario of progressive, slow incision of the African Surface and concomitant water table lowering.

Major and trace element data, determined by XRF, are shown in Table 1. Although parent material and hence saprolite values are very variable, progressive leaching of the more mobile elements, for example Na, Ca, Mg and K, is evident in the saprolite, these elements being almost entirely leached out of the residuum where there is relative accumulation of the less mobile elements.

The profile feature which most strongly indicates the nature of the operative processes, as concerns landscape development, is the configuration of the boundary between saprolite and residuum. Its diffuse form is typical of a dissolution front, where saprolite leaching is sufficient to cause the texture to collapse. In the upper parts of profiles exposed in road cuttings, similar features are exposed and where quartz stringers occur in the saprolite their collapse results in local occurrences of

Table 1. *Major and trace elements (XRF) in the Chimimbe interfluve profile (B3)*

Depth (m)	SiO$_2$ (%)	Fe$_2$O$_3$ (%)	Al$_2$O$_3$ (%)	TiO$_2$ (%)	MnO (%)	MgO (%)	CaO (%)	Na$_2$O (%)	K$_2$O (%)	P$_2$O$_5$ (%)	Ba (ppm)	Cr (ppm)	Mn (ppm)	Nb (ppm)	Ni (ppm)	Rb (ppm)	Sr (ppm)	V (ppm)	Y (ppm)	Zr[1] (ppm)
Residuum																				
1.0	65.61	10.33	21.48	2.20	0.099	0.13	0.06	0.03	0.162	0.15	78	100	742	16.0	61	19	15	222	29	463
2.5	58.46	13.32	25.01	2.37	0.101	0.12	0.06	0.09	0.149	0.10	91	151	666	16.5	61	11	13	257	40	470
5.5	42.71	28.10	25.72	2.03	0.525	0.13	0.06	0.02	0.116	0.10	557	349	2990	13.2	56	11	11	488	43	321
6.0	45.34	31.34	17.01	2.29	1.620	0.09	0.04	0.01	0.052	0.28	2302	598	10647	14.4	81	5	6	612	58	227
7.0	43.05	24.11	27.55	3.90	0.152	0.27	0.10	<0.01	0.104	0.07	219	100	934	13.3	62	11	12	400	41	266
Saprolite																				
8.3	70.54	2.28	25.08	0.35	0.014	0.72	0.05	0.02	1.050	<0.005	280	56	90	1.5	43	112	36	39	12	47
10.0	52.10	13.57	29.87	1.19	0.246	1.15	0.34	0.10	0.632	0.01	381	180	1369	4.9	97	73	31	133	54	82
12.5	44.39	19.97	24.43	2.26	0.127	2.87	3.05	0.41	0.461	0.02	199	475	669	5.2	94	8	36	287	54	99
13.0	51.69	7.47	38.44	0.59	0.048	0.33	0.33	0.01	0.091	0.01	362	284	269	1.1	154	4	23	53	19	59
13.5	50.91	9.00	33.96	0.69	0.079	1.97	2.01	0.28	0.328	0.02	498	292	439	0.8	181	6	33	60	26	39
15.5	56.10	13.47	17.07	1.71	0.195	3.34	2.79	2.75	1.990	0.52	652	234	1263	10.7	64	<4	280	223	59	183
Rock																				
20.0	57.55	5.42	18.48	0.42	0.090	4.12	8.35	5.20	1.050	0.07	225	608	618	2.4	44	10	471	96	17	59
20.5	48.13	15.85	14.31	2.09	0.200	6.44	10.06	2.79	1.110	0.26	148	123	1377	8.0	49	12	201	366	55	143
21.0	73.09	1.19	16.88	0.18	0.030	0.32	3.53	5.74	0.940	0.05	290	4	181	0.7	6	16	459	21	4	77

stonelines at the boundary. The wider occurrence of these stonelines in this area is described elsewhere (McFarlane & Pollard 1989). A measure of the local variation in the position of this dissolution front is provided by the gamma logs of two holes drilled closely adjacent to B3. These were percussion-drilled, but by means of correlating the details of the profile stratigraphy and gamma 'signatures' in the B3 core, it becomes possible to assign profile boundaries to the percussion hole profiles purely on the basis of the gamma logs, as shown in Fig. 6. The variable depth of the saprolite/residuum interface in these holes provides a measure of the local scale of the subsurface relief on this basined interface, within the wider scenario of larger-scale basining reflected in the topography (McFarlane & Pollard 1989).

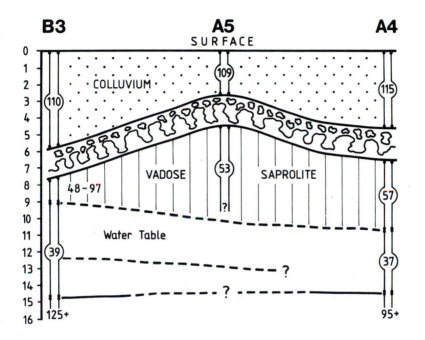

Fig. 6. Gamma 'signatures' (100 KeV means) of horizons of the interfluve profile (B3) and inferred horizons in adjacent boreholes A5 and A4. (No vertical exaggeration on the diagram.)

From the mean values of elements in the saprolite and residuum, the concentration factors for individual elements in the conversion of saprolite to residuum can be calculated (Table 2). Thus the percentage losses of the residual elements in the residuum can be assessed against the element with the greatest concentration factor, in this case Mn, with Zr as a close second best. Although SiO_2, Al_2O_3, Fe_2O_3 and TiO_2 comprise over 98% of the residuum, the conversion of saprolite to residuum results in losses of 88%, 80%, 54% and 46% respectively and a minimum of about 30 m of surface lowering is deduced (McFarlane 1988a, McFarlane et al. in prep.). The extensive loss of Al_2O_3 is consistent with congruent kaolinite dissolution, this being essentially responsible for the collapse of the saprolite. The profile chemistry thus supports a model of land surface formation involving differential leaching (or etchplanation) rather than mechanical stripping (or pediplanation).

Table 2. *Element concentration factors and percent losses in the conversion of saprolite to residuum*

	Concentration factor	% lost		Concentration factor	% lost
SiO_2	× 1.102	88.0	Nb	× 4.277	14.30
Fe_2O_3	× 2.297	54.0	Ni	× 0.715	85.70
Al_2O_3	× 0.970	80.6	Rb	× 0.524	89.45
TiO_2	× 2.689	46.2	Sr	× 0.847	83.05
MnO	× 4.992	—	Y	× 1.327	73.43
CaO	× 0.052	99.0	Zr	× 4.840	3.08
MgO	× 0.100	98.0	Ba	× 1.924	61.47
Na_2O	× 0.064	98.7	Cr	× 1.200	75.97
K_2O	× 0.180	96.4	V	× 3.507	29.78
P_2O_5	× 1.685	66.3			

The chemistry of this profile also allows characterization of water associated with the different horizons. Water in the vadose zone must inevitably be poor in bases since these are released into the groundwater in the zone of inferior alteration in the lower part of the profile. The upper 6 m of the saturated zone must also have base-poor water because this is the zone of 'superior alteration' which was formerly vadose. Shallow water passing through the residuum, that is water from 3–7 m depth could be expected to carry metallic elements released during the advanced stages of leaching associated with saprolite collapse, e.g. Al, Fe and Ti.

Linthembwe interfluve profiles (DW67 and BHAH9). Profile data is only available from percussion-drilled holes on the interfluves of the Linthembwe catena. On the northeast interfluve, first water strike in DW67 was at 15 m at the transition from reddish brown to greenish grey colouration, with the final rest level at 12.2 m. In BHAH9 on the southwest interfluve, first strike was at 10.2 m, with a rest level of 5.8 m. This is consistent with the observation at Chimimbe that the present water table is relatively high, in the zone of superior alteration; at Linthembwe it appears that some 2.8–4.4 m of former vadose saprolite is now below the water table.

Dambo and dambo-peripheral profiles

Textural details of the infill of Linthembwe and Chimimbe dambos were provided from Tripple Core Barrel and Cobra Piston cores, Minute Man percussion-drilled holes and pits.

Chimimbe dambo. The general configuration of the interface between infill and underlying saprolite is shown in Fig. 7. Gravels were not encountered at the lowest parts of the depressions. Nor did sand occur consistently at the base. In P36, for example, 1 m of sandy clay overlay very sandy clay, with clay as the basal component directly overlying gneissic saprolite. Textural variations were generally inconsistent with the 'fining upwards' hypothesis and the cores and pits showed that the quartz occurs as unsorted float within the clay. In a core from the dambo centre, which had

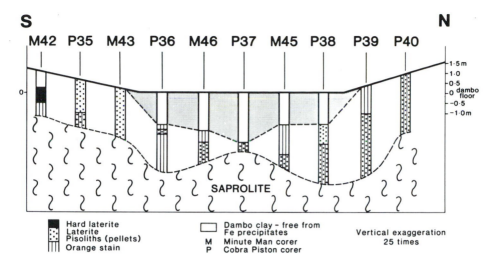

Fig. 7. The general configuration of the interface between infill and underlying saprolite below Chimimbe dambo. Lateritic precipitation is most pronouned in the clay beneath surficial sand in marginal situations, but also occurs in the lower parts of the central infill (from logs by E. Freshney).

clay as the basal component, the boundary with the saprolite was highly convoluted (Fig. 8). At the base of this and other dambos, the logs commonly record clasts of saprolitic material within the lower part of the clayey infill and discrete masses of clay within the upper part of the saprolite. It is difficult to reconcile these stratigraphic and boundary features with a fluviatile sequence. The nature of the interface between infill and saprolite is very similar to the interface between residuum and saprolite in interfluve situations. On the gently sloping margins of the dambo, although the surface layers become sandier, grey clay with quartz float, in appearance entirely similar to the contemporary dambo clay, underlies this. The clay below the sand at the dambo margins is impregnated by lateritic precipitates (Fig. 7) which prevented penetration sufficient to establish the existence of an erosional boundary between infill and the regolithic profile and in particular failed to identify a diagnostic 'wedge' of residuum or colluvium (Fig. 2) below the periphery of the dambo sediments, as hypothesized by Freshney (1985, 1987).

XRD of the clay plus silt fraction from profile P36, drilled on the dambo floor, showed (Table 3) that smectites dominate, with kaolinite in smaller proportion. In P33 and P40, both at the dambo periphery, smectites have been leached out of the near-surface horizons, leaving sandy kaolinitic material. The underlying grey clays, which are texturally similar to the dambo clays, show increasing proportions of smectite. This stratigraphy is consistent with a leaching sequence in which smectite is replaced by kaolinite and, much as in the case of interfluve situations, congruent kaolinite dissolution yields residual surface sands. These sands, the so-called 'wash sands' (Mackel 1974, 1985), exposed in numerous shallow, hand-dug wells around the dambo, are conspicuously unsorted and lack stratification.

Where it was possible to examine the nature of the interface between surficial material and saprolite in the dambo peripheral area (McFarlane & Pollard 1989), the configuration was irregular and consistent with a dissolution front, as in both the

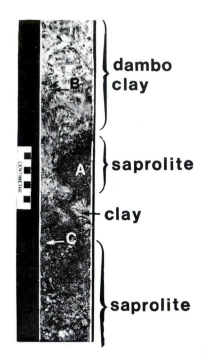

Fig. 8. Detail of the interface between dambo clay and underlying saprolite in a core from Chimimbe dambo. A clast of saprolite (B) occurs suspended within the clay above the interface. A near-clast of saprolite (A) is only linked with the saprolite by a thin 'neck'. An invasion of dambo clay occurs within the main body of the saprolite (C).

Table 3. *X-ray diffraction results of clay plus silt fractions of the Chimimbe dambo and dambo-peripheral sediments. Location of P40 and P36 are shown in Fig. 7. P33, not shown, is located about 60 m upslope from M42.*

Profile	Depth (m)	Smectite (%)	Kaolinite (%)	Quartz (%)
P33	1.0	—	62	38
	1.8	5	79	16
(Dambo periphery—southside)	3.5	38	59	2
	3.7	36	62	2
P36	0.8	57	25	18
	1.8	71	14	15
(Dambo floor)	3.9	65	18	17
P40	0.8	—	87	13
	1.8	—	73	27
(Dambo periphery—north side)	2.7	34	56	10
	2.9	43	53	4

cases of the interfluve and dambo floor profiles. As in the case of the logs of holes drilled on the dambo floor, those drilled on the flanks also recorded clay masses within the saprolite (e.g. P35 in Table 4), clasts of saprolite within the lower part of the clay (e.g. P33) and what is evidently a very diffuse, convoluted boundary, as shown for example by P34.

Table 4. *Examples of the nature of the interface between sediments and saprolite (from logs by E. Freshney). Location of P35 is shown in Fig. 7. P34 and P33, not shown, lie about 30 m and 60 m upslope from M42.*

P35	0.70–1.10 m	Hard, very sandy clay with laterite
	1.65–2.05 m	Clay, grey, very sandy with some orange stain and some dark ferromanganiferous nodules. Abundant quartzo-feldspathic debris at base
	2.35–2.75 m	40mm of quartzo-feldspathic material on dark clayey *rotten gneiss with some clay masses*
P33	0.65–0.80 m	Sand, coarse grained, clean, some grains subrounded
	0.80–1.05 m	Sand, very clayey, coarse grained. Clay content rising down to extremely sandy clay. Some lateritic material in lower part
	1.65–2.05 m	Grey clay, much solid laterite especially bottom 0.05
	2.45–2.65 m	Clay, grey, sandy, with laterite
	2.65–2.85 m	Sand, clayey, grey and yellowish brown with some masses of grey clay
	3.50–3.60 m	Sand, buff, medium grained
	3.60–3.80 m	*Clay* extremely sandy grey and orange with some green patches. *Some masses of soft rock*
	3.80–3.90 m	? Gneiss, dark brown rotten
P34	0.60–1.00 m	*Sand*, clayey medium grained at top but coarse down. Some laterite near base
	1.65–2.05 m	*Clay*, grey with much ferruginous material and ? *masses of rotten rock*
	2.25–2.40 m	Dark brown, *very rotten rock*
	2.40–2.65 m	*Clay*, pale grey with *small rotten rock masses* and some ? feldspar. Dark clay patch at 2.6
	3.15–3.55 m	Dark brown, *very soft rock*, possibly basic gneiss. Some feldspathic material. *Patch of greenish grey clay* with feldspathic material 3.35–3.45
	3.55–3.85 m	*Rotten gneiss*, brown, with some ? garnets

In short, the dambo and dambo-peripheral profiles did not show stratigraphic and boundary features consistent with fluviatile erosion and sedimentation. Moreover, the absence of smectite on the interfluves precludes the possibility that the smectite in dambos is transported from higher catenary positions.

Linthembwe dambo. In this dambo, narrower and more deeply inset into the terrain, the infill is considerably thicker. The general configuration of the interface with the underlying saprolite is shown in Fig. 9. As at Chimimbe, textural variations, logged

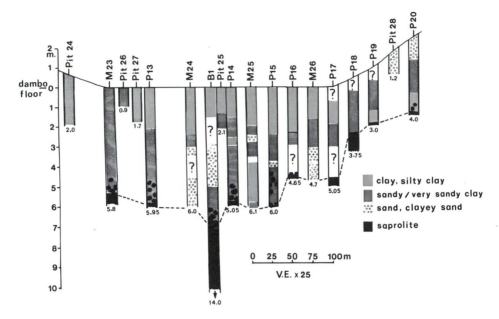

Fig. 9. The textural variation of the infill of Linthembwe dambo and general configuration of the interface with the underlying saprolite (from logs by E. Freshney).

from drilled holes and dug pits on the flat dambo floor failed to show a 'fining upwards' sequence (Fig. 10). Similarly, the quartz sand, some of which is rounded and some sharply angular, occurred as 'float' within the clay. Interestingly, when the lower, sandy, part of the infill in one of the cores dried out, magnesium sulphate crystallization indicated some rudimentary stratification (? suffosion). The basal component was in this case (Fig. 10) a sharply angular pegmatitic rubble which protruded steeply upwards into the sand on one side of the core. X-ray diffraction again identified smectite, kaolinite and quartz (Fig. 10). Neither smectite nor kaolinite was identified in the underlying saprolite which was dominated by dis-aggregated primary minerals. The upper part of the hard rock showed incipient penetration of weathering particularly of the micaceous partings and along fractures (saprock). As at Chimimbe, poor penetration in dambo peripheral situations pre-vented the identification of an erosional boundary between infill and regolith and a trench dug across the hypothetical site of the boundary on the north side of the dambo failed to find it (Freshney 1987). Beyond the confines of the flat valley floor, as at Chimimbe, the profiles have sandy surface residua overlying kaolinitic clay and with smectites preserved at deeper levels in material which is texturally similar to the modern dambo clay.

A core drilled about 11 m above the dambo floor on the southeast side (B2 in Fig. 3) provided strong support for the contention that the dambo is flanked by former dambo clay infill presently being modified by leaching. The profile is shown in Fig. 11. Some 3 m of red, sandy clay (*c.* 10% kaolinite at 2 m) overlay about 7.5 m of paler, greyish yellow sandy clay, with sand content increasing up-profile (*c.* 3% kaolinite at 4 m). Lateritic precipitation occurred in the lower, more clayey part. As is common in other slope-bottom laterites, the vermiform cavities of the laterite encapsulated pale grey clay with suspended quartz, similar in colour and texture to

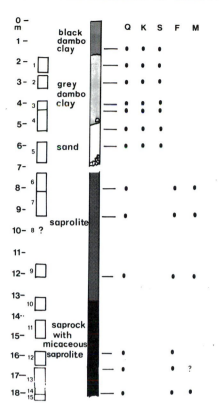

Fig. 10. Main features of a core from the centre of Linthembwe dambo. (Recovered core sections are boxed and numbered.) Q, quartz; K, kaolinite; S, smectite; F, feldspar; M, mica.

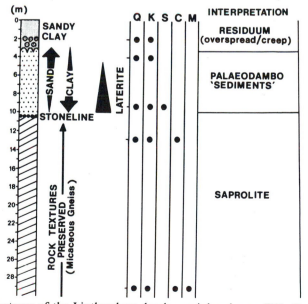

Fig. 11. Main features of the Linthembwe dambo-peripheral core (B2). Q, quartz; K, kaolinite; S, smectite; G, goethite; C, chlorite; M, mica.

the modern dambo infill. It is significant that smectite was not identified in the underlying micaceous saprolite but occurred within the vermiform cavities. This profile is thus interpreted as a 'palaeodambo' clay, now over-run by interfluve-type colluvium or residuum (McFarlane *et al.* in prep.).

Catenary stratigraphy

Diagrammatic synthesis of the stratigraphy of Chimimbe catena is shown in Fig. 12. The broader configuration of the basal surface of weathering follows the relief of the terrain, but is generally more subdued. Saprolite is thickest where micaceous rocks occur, in the dambo periphery. Saprolite is thinnest below the dambo itself. The dambo clay is stratigraphically continuous with palaeodambo clay on the flanks of the dambo below surficial sands. In the lower parts of the clay there is lateritic precipitation of Fe leached from the upper part of the interfluve profiles.

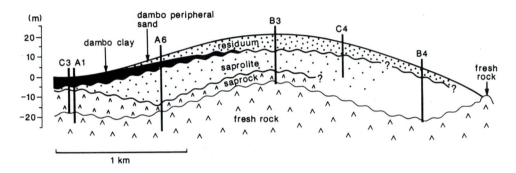

Fig. 12. Diagrammatic synthesis of the stratigraphy of Chimimbe catena.

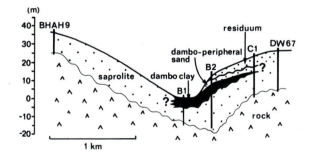

Fig. 13. Diagrammatic synthesis of the stratigraphy of Linthembwe catena.

The stratigraphy of the Linthembwe catena is summarized in Fig. 13. Again, the saprolite is deepest where micaceous rocks occur along the dambo flank and thinnest below the dambo clay. In this more deeply inset dambo the smectitic infill is much thicker than at Chimimbe.

In summary, the catenary stratigraphy is consistent with the genetic model involving differential leaching, saprolite collapse and landsurface lowering, with the dambos occupying the 'lows' of an irregularly lowered surface. It also provides support for the deduction, based on the evolution of dambo morphology, that the dambos are reduced in area and inset within the original landsurface, leaving a palaeodambo clay wedge along their flanks. No evidence was found to support the hypothesis that the dambos are of fluviatile origin. Hence the hypothetical presentation in Fig. 2b, rather than that presented in Fig. 2a, appears to apply.

Water movements

Catenary variations in permeabilities in representative profiles at Chimimbe are shown in Fig. 14. Essential features are the up-profile reduction in permeability in the interfluve profile and, in a lateral direction, the progressive reduction in permeability of surficial material, from interfluve residuum, through palaeodambo clay to dambo clay. This pattern is consistent with the stratigraphic/genetic model. It implies that recharge to the deep water system is at a maximum on the interfluves and that in progressively lower catenary positions lateral movement of infiltrating water, as shallow throughflow, would be favoured, discharge occurring at the dambo margins where the surficial sands feather out. There is nothing to suggest recharge of the deep groundwater directly through the dambo floor, although such recharge can be deduced where dambos occur as discrete, karst-like depressions in elevated situations (McFarlane & Pollard 1989).

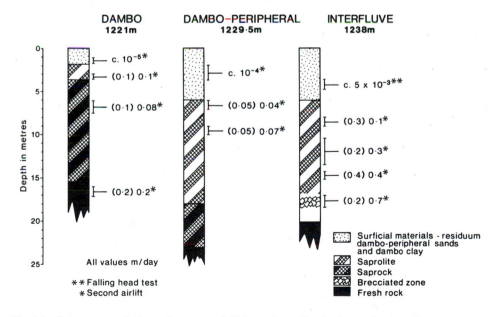

Fig. 14. Catenary variations in permeabilities of profile horizons in interfluve, dambo-peripheral and dambo situations, at Chimimbe.

Data on potential groundwater movements in the dambos are presented in Fig. 15. In the wet season, heads are consistently such as to promote upward discharge of deep water through the dambo clay. This is the case for the more deeply inset

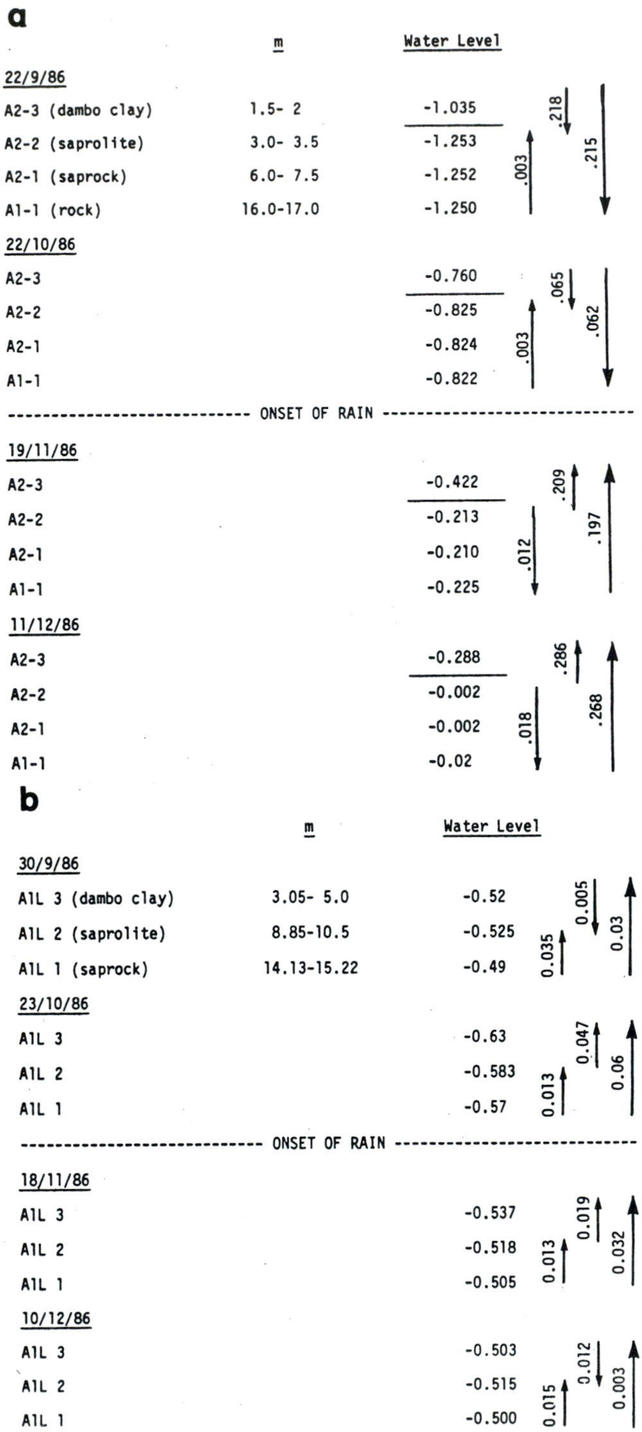

a

	m	Water Level

22/9/86

A2-3 (dambo clay)	1.5- 2	-1.035
A2-2 (saprolite)	3.0- 3.5	-1.253
A2-1 (saprock)	6.0- 7.5	-1.252
A1-1 (rock)	16.0-17.0	-1.250

.218 .003 .215

22/10/86

A2-3		-0.760
A2-2		-0.825
A2-1		-0.824
A1-1		-0.822

.065 .003 .062

------------------------- ONSET OF RAIN -------------------------

19/11/86

A2-3		-0.422
A2-2		-0.213
A2-1		-0.210
A1-1		-0.225

.209 .012 .197

11/12/86

A2-3		-0.288
A2-2		-0.002
A2-1		-0.002
A1-1		-0.02

.286 .018 .268

b

	m	Water Level

30/9/86

A1L 3 (dambo clay)	3.05- 5.0	-0.52
A1L 2 (saprolite)	8.85-10.5	-0.525
A1L 1 (saprock)	14.13-15.22	-0.49

.005 .035 .03

23/10/86

A1L 3		-0.63
A1L 2		-0.583
A1L 1		-0.57

.047 .013 .06

------------------------- ONSET OF RAIN -------------------------

18/11/86

A1L 3		-0.537
A1L 2		-0.518
A1L 1		-0.505

.019 .013 .032

10/12/86

A1L 3		-0.503
A1L 2		-0.515
A1L 1		-0.500

.012 .015 .003

Fig. 15. Water levels and heads in dambo floor profiles, from nests of piezometers emplaced within the clayey infill and the underlying saprolite and rock. (a) Chimimbe dambo; (b) Linthembwe dambo.

Linthembwe dambo even in the dry season. However, field observations, in the dry season, show that the dambo clay is effectively impermeable, preventing upward discharge through it. As illustrated in Fig. 16, in the central parts of Linthembwe dambo the grass was dead, the clay deeply cracked, and there were no magnesium sulphate efflorescences from the sulphate-rich groundwater. Towards the seepage zone at the dambo margin, grass growth was sustained, clay cracking became less pronounced and magnesium sulphate efflorescences occurred. In the dry, central part of the dambo circles of uncracked clay and green grass around one of the boreholes and around an isolated, woody shrub, presumably with roots sufficiently deep to penetrate the clay 'lid', confirmed that although there is potential for upward movement of groundwater this does not generally occur unless the 'lid' is breached.

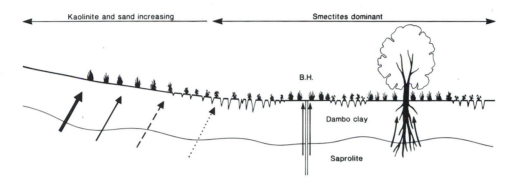

Fig. 16. Schematic transect of Linthembwe dambo to show the distribution of green or dead grass and the cracking of the smectitic clay. For explanation see text.

Water movements may be summarized as shown in Fig. 17. In effect, the palaeo-dambo clay wedge separates shallow and deep water discharge towards the dambos. The linear continuity of the seepage zone, well seen as a dark-toned strip on air photos taken in the dry season, indicates continuous discharge of shallow through-flow. Upward discharge of deep water is predominantly near the margins of the dambos, at or in the vicinity of the seepage zone. Upward discharge in more central parts is evidently frustrated and lateral movement 'downstream' below the dambo clay 'lid' deduced. Such under-the-'lid' movement is consistent with the linearity of the narrower, more deeply inset dambos, which clearly express geological control. A lithological orientation could not be brought about by the direct surface runoff, in the wet season, along the dambo floors since such water is separated from the saprolite by the dambo clay infill; only the movement of water within the saprolite below the clay could allow dambo orientation to express the geology. The leaching activity (and suffosion?) of this under-the-'lid' water movement is held to be responsible for the progressive shrinking and insetting of the dambos within the African Surface terrain and for the thinning of the saprolite under the dambos. The invasion of the dambo system first by gullies and then by stream systems appears to be associated with thinning of the sub-clay saprolite to the extent that wet season groundwater movement within it is so constrained that it breaks through the 'lid'.

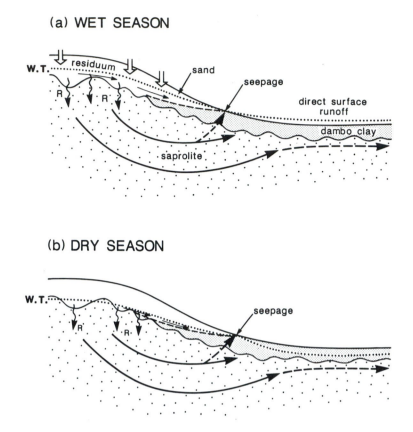

Fig. 17. Schematic representation of water movements associated with a dambo catena. (a) Wet season precipitation which infiltrates the interfluve profile has a shallow throughflow component ('rapid response') which discharges onto the dambo floor at the seepage zone. This joins the direct surface runoff from the dambo floor after the smectites have taken up water sufficient to seal the cracks. Recharge to the deep system occurs through the basined saprolite/residuum interface. Upward discharge of deep water occurs at the dambo margin and as discrete springs on the dambo floor. Elsewhere across the dambo, the clay 'lid' frustrates discharge, diverting the deep water laterally 'downstream' below it. (b) Slow recharge to the deep groundwater continues through the basined boundary of the interfluve profile. Shallow throughflow is reduced, but continues to discharge at the dambo margins, where it is largely lost by evaporation. Upward discharge at springs and lateral movement of deep water below the dambo clay continues, becoming progressively less vigorous as the head within the interfluve is reduced.

Water quality

Natural discharges

The catenary stratigraphy, which controls the water movements, carries clear implications as concerns the chemistry of these shallow and deep waters, which may be expected to be significantly different since they pass through parts of the profiles with mineral assemblages pertaining to different stages in the leaching progression. Thus, in general terms, deep water would be expected to be rich in bases. Water from shallow sources would be poor in these and could be expected to carry elements released during the advanced stages of weathering.

Seepage zone water. A series of 11 shallow pits, about 30 m apart, was dug on the transect upslope from Chimimbe seepage zone. In the lowest of these, the water level was closest to the surface, at a depth of 0.44 m, becoming progressively deeper in an upslope direction, where a level of 1.22 m was recorded in the highest pit, some 8 m above the dambo floor. Electrical conductivity (Allen 1988) ranged from 56 to 150 µs/cm. These low values express low solute load and reflect laterally downslope diversion of water after its infiltration and passage only through the sandy kaolinized upper parts of the profiles. This presents a clear contrast with the high EC values of the waters sampled from holes drilled into the dambo floor (Fig. 18).

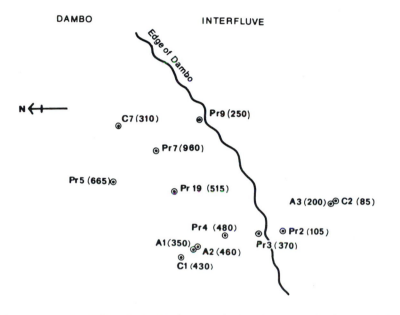

Fig. 18. Electrical conductivities (µs/cm) of waters in boreholes on the floor and flanks of Chimimbe dambo. The Proline holes (Pr) were about 5 m deep and all other holes 20–30 m deep (after Shedlock 1990).

Although the lateral continuity of the seepage zone, even in the dry season, indicates sustained discharge of shallow throughflow, upward discharge of deep water is implied by the otherwise inexplicable, very wet patches which occur locally along it. Seepage water was sampled at two particularly wet localities at the edge of the Linthembwe dambo at the end of the dry season, following a poor wet season. The samples were filtered (0.45 µm) in the field and acidified and cool-stored in the UK prior to analysis. ICP–OES data (Table 5a) confirm that in both cases there is a deep water component expressed by the bases. A shallow throughflow component is weakly indicated in the case of the first sample which has low Si and relatively high Fe and Mn values. Al was below the detection limit (<0.1 ppm).

Table 5. *Chemical analyses (ppm) of seepage zone waters (a) and crescent springs on a dambo floor (b), analysed by ICP–OES*

	Field pH	Na	Ca	Mg	K	Cl	Si	Al	Fe (tot.)	Mn	Ni	SO_4
(a)												
1	5.9	64	95	57	1.7	4	10	—	1.22	2.50	0.07	368
2	6.6	71	114	104	1.6	16	23	—	0.03	0.27	—	639
(b)												
3	6.4	28	62	16	—	2	27	—	0.13	0.40	—	114
4	7.0	163	555	343	0.8	5	24	—	0.23	0.16	—	2490
5	6.9	67	143	78	1.4	19	25	—	0.20	0.10	—	528
6	6.7	81	391	108	0.8	6	41	—	0.02	0.15	—	1240

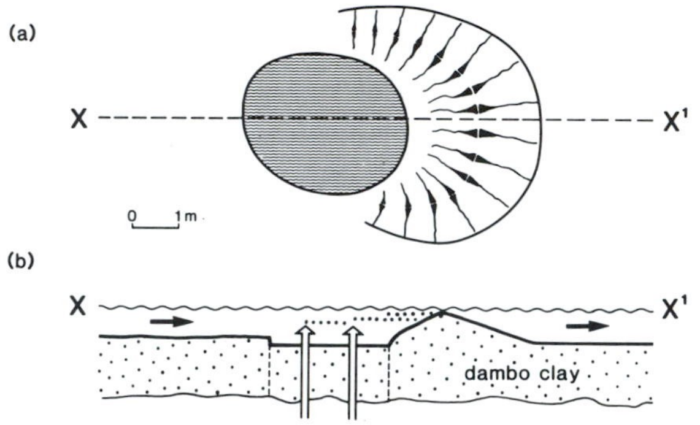

Fig. 19. Diagrammatic representation of a crescent spring, showing the plan (a) and transect (b).

Crescent springs. Upward discharge of deep water also occurs at discrete springs along the margins of the dambo floors, below the seepage zone. Where discharge is strong, the springs develop a distinctive crescentic form (Fig. 19) deduced to result from the deposition, by direct surface runoff in the wet season, of clay disturbed by upward discharge. The waters of a selection of such springs, aligned on both sides of a 'tributary' of Linthembwe dambo (Fig. 20) were sampled and analysed by ICP–OES (McFarlane *et al.* 1988). As with the seepage water, the samples were field filtered (0.45 μn) and acidified and cool-stored in the UK prior to analysis. Results are shown in Table 5b. The waters have remarkably varied chemistry, ranging from long-residence deep water, rich in bases (e.g. samples 4 and 6) to shallow water which is base-poor (sample 3). These striking variations within short distances (*c.* 30 m) are consistent with discharge from fractures drawing water from different horizons in the weathering profile as illustrated in Fig. 21. The importance of by-passes (fractures and stringers) within the regolith, as also suggested by other research (e.g. McFarlane *et al.*, this volume) is clear. The progressively lower water levels in the springs in a downstream direction indicate that the fracture orientation is transverse to the tributary dambo, parallel with the linearity of the main dambo (Fig. 22).

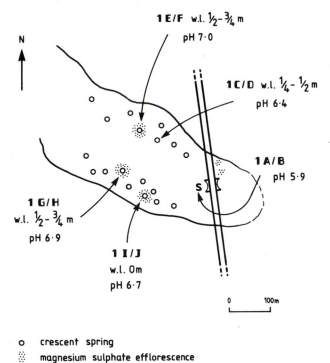

Fig. 20. Location of a group of crescent springs in a tributary of Linthembwe dambo, selectively sampled for analysis by ICP–OES (see Table 5b).

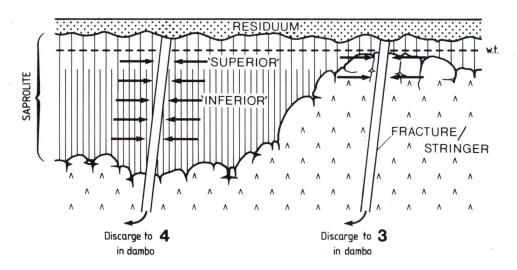

Fig. 21. Variations in the proportions of 'superior' and 'inferior' saprolite in profiles can account for fracture-related discharges with very different chemical compositions, as in the case of the crescent springs.

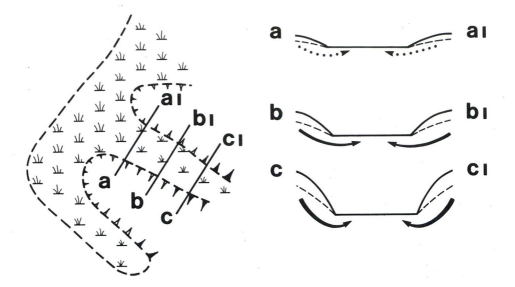

Fig. 22. Progressively lower water levels in the crescent springs, in a 'downstream' direction, would be consistent with discharge from fractures parallel with the main dambo, since the heads would decline as the altitude of the local spurs lowers towards it.

A high proportion of bases to silica characterizes water derived from the zone of 'inferior' weathering. In these and other spring waters analysed (McFarlane *et al.* 1988), elevated sulphate values were associated with high base to Si ratios, this indicating that the source of the sulphate is deeply weathered profiles with thick saturated zones and not, as was earlier believed to be the case, in situations in which sulphide-bearing rocks protrude upwards into the vadose zone where oxidizing conditions occur. The link between high sulphate values and long-residence deep water is also indicated by close correlations with high EC values, reflective of total dissolved solids (Kafundu 1984).

Relatively high chloride values are associated with water which has 'deep' chemical signatures, the lowest value (2 ppm) being associated with the water which is poorest in bases. This appears to express little evapotranspirative loss from the upper parts of the water body within the regolith profiles. The relatively high values in long-residence discharge imply derivation from deeper pockets of weathering in which near-stagnant water has facilitated chloride concentration from the parent rocks. These patterns suggest that the use of chloride in river water to estimate evapotranspirative losses should be approached with caution.

The apparent absence of Al in the water from shallow sources is conspicuous and is inconsistent with the deduction, based on the chemistry of the interfluve profile, that aluminium is released into shallow throughflow following congruent kaolinite dissolution.

Well water. The rural people draw water predominantly from shallow (1–2 m) hand-dug wells at the margins of the dambos. Some of these wells would be expected only to draw on shallow throughflow, water moving laterally downslope over the wedge of palaeodambo clay. Some would be expected to have such water augmented by upward discharge of deep water. Seven wells were sampled near the margins of Chimimbe and Linthembwe dambos. They were filtered in the field and acidified and cool-stored in the UK till analysed by ICP–OES. Results are shown in Table 6. Low EC waters, those free from or only weakly augmented by deep water, are low in bases and silica. Values rise with EC, indicating augmentation by deeper water. SO_4 and Sr values are also higher where EC and bases are high. Conversely, Fe is lowest where the EC is greatest. Aluminium is below the detection limit of 0.1 ppm, even in low EC, shallow water. This is inconsistent with the deduction that it is leached out during congruent kaolinite dissolution in the higher parts of the interfluve profiles and also with the deduction that the palaeodambo sandy residuum results from similar dissolution.

Table 6. *Chemical analyses (ppm) of shallow well waters near the margins of Chimimbe and Linthembwe dambos (analysed by ICP–OES).*

| | Chimimbe Dambo | | | | Linthembwe Dambo | | |
	F22	F25	F26	F30	F4	F15	F17
EC	59	100	120	230	280	990	150
pH	6.63	7.40	7.31	7.85	5.94	6.24	6.27
Na	2.8	3.8	3.7	10.2	4.3	78.1	16.9
K	1.1	<0.8	2.1	1.6	<0.8	15.3	2.2
Ca	4.7	10.5	8.6	30.8	0.6	108.0	13.9
Mg	2.1	3.7	5.3	13.1	0.3	62.6	7.4
B	<0.03	<0.03	<0.03	0.03	<0.03	0.06	0.05
Li	<0.007	<0.007	<0.007	<0.007	<0.007	<0.007	<0.007
SO_4	<0.8	<0.8	<0.8	<0.8	1.8	581.0	20.7
Si	10.4	9.6	10.8	35.6	13.7	24.5	6.7
Cr	<0.04	<0.04	<0.04	<0.04	<0.04	<0.04	<0.04
Sr	0.050	0.081	0.064	0.309	0.007	0.967	0.150
Ba	0.061	0.059	0.122	0.157	0.023	0.085	0.145
Y	<0.003	<0.003	<0.003	<0.003	<0.003	<0.003	<0.003
Mn	0.053	0.013	0.116	0.030	0.014	0.137	0.532
Fe	0.077	<0.015	<0.015	0.366	0.125	<0.015	0.016
Zn	<0.020	0.106	<0.020	0.412	0.035	<0.020	0.020
Al	<0.10	<0.10	<0.10	<0.10	<0.10	<0.10	<0.10

Evaporites. In the dry season, when there is no direct surface runoff from the dambos, discharging water evaporates. Chemical and biochemical precipitates include magnesium sulphate efflorescences which commonly encrust grass stems (McFarlane 1986), gypsum and calcrete nodules within the upper part of the dambo clay profile and Fe and Mn lateritic precipitates particularly at the dambo margins. Thus the dambos are sites where there is neoformation of minerals, that is direct formation from materials carried in the waters as opposed to secondary mineral formation which replaces earlier phases during the differential leaching process. These precipitates provide a means of identifying what is in the discharging waters, other than by direct analysis of the waters themselves.

Less conspicuous than the magnesium sulphate crystallization, but very much more widespread, particularly in the headwater reaches and dambo margins, is the formation of a wide variety of alumino-silicate precipitates. These occur as tiny, grey-coloured glaebules also encrusting grass stems (McFarlane 1988*b*). Morphology of the individual particles comprising the glaebules is highly varied, ranging from sharply tabular crystals (quartz or crystobalite), through poorly crystalline materials (with widely variable silica/alumina ratios) to framboid-like bodies (almost entirely aluminous), gelatinous material and pseudomorphs after fungal hyphae and spores (high proportion of Si in relation to Al). Other elements in the representative selection of particles analysed (using a Jeol JXA-840 scanning electron microscope coupled to a Link AN10000 analyzer) ranged as shown in Fig. 23. Thus, both base-carrying deep water and shallow water carrying Al and Ti are represented in these evaporites. In effect, everything leached from all parts of the interfluve profiles is reassembled in these precipitates on the dambo floors. This material appears, on progressive burial, to evolve through a non-expanding 2:1 silicate phase identified in surface horizons (Bird *et al.* 1988) to the smectite and kaolinite assemblage characterizing the bulk of the dambo clay infill.

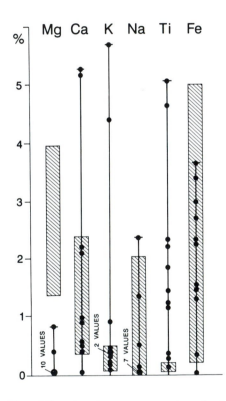

Fig. 23. Mg, Ca, K, Na, Ti and Fe values in a representative selection of particles comprising the alumino-silicate evaporites, which occur as glaebules encrusting grass stems on dambo floors. Stippled columns represent ranges of values for montmorillonite published in Grim (1968).

Discussion

From the catenary variations in the profile it is clear that in addition to upward discharge of deep water into the dambos via fractures, shallow throughflow moves laterally downslope, passing over the palaeodambo clay wedge at the dambo margins and discharging at the seepage zone. These are quite distinctive waters and although this is expressed in terms of lower EC, bases and Si and higher Fe in the shallow component, nevertheless the deduced Al content of such water is apparently absent. Aluminium is lost from the interfluve profiles, as shown by the mass balance study of interfluve core B3. The evaporites show that it arrives in the dambos in the dry season. Yet it was not detected in the seepage waters and shallow throughflow which were also collected in the dry season. This inconsistency could be debated in terms of 'at the time the water was collected the aluminium was not mobilized in the water; perhaps mobilization pertains to the early part of the dry season, not the late?'.

However, a similar problem was encountered where such rationalization could not be applied; in microbial leaching experiments, using micro-organisms indigenous to profiles in which congruent kaolinite dissolution is occurring, the aluminium levels detected in the leachates were also inappropriately low. In the first in-flask experimental work (McFarlane & Heydeman 1984), kaolinitic substrate and leachate were separated by centrifugation and although chemical analysis of substrate showed substantial loss of Al, in the leachates, which were analysed by a spectrophotometric method, Al values were inappropriately low. Re-analysis by AAS (atomic absorption spectrometry) also failed to detect Al in appropriate quantity (Hughes, pers. comm.). The experiment was repeated and the leachates analysed by FAB (fast atom bombardment), which also failed to detect aluminium (Hotten, pers. comm.). Obviously if the aluminium is missing from the substrate it *must* be in the leachate. Yet three analytical procedures in the case of the experimental work only established its occurrence in inappropriately small quantities or not at all. In natural waters with even lower expected concentrations the failure of ICP–OES to detect it, even though it could be expected at values well over instrumental detection limit, clearly indicates that there is a fundamental problem in aluminium determination.

Because a range of techniques was used, it can be deduced that the problem does not lie with the analytical techniques but with the pre-analytical treatment of the samples, all of which were filtered and acidified. This poses a serious question: is the methodology for the assessment of elements in solution appropriate to the assessment of Al levels in natural groundwaters? From this stems concern over the validity of the determination of other metallic elements in such water. For example Fe and Cr were also deduced to have been leached from the Malawian interfluve profile and like Al, they would not be expected to go into solution under the ambient conditions. However, if, as is evidently the case with Al, solution is not synonymous with mobilization, then further systematic research is clearly needed on pre-analytical treatment of tropical waters. Since the rural people are drinking untreated water, it is the total chemical content which should be of concern rather than the results of analyses of filtered and acidified water.

Conclusions

A systematic examination of the variations in profiles in catenary relationships, placed in geomorphological context, can contribute to the understanding of water circulation systems and water chemistry of low-relief terrain in areas of basement rocks. This is because the deeply weathered and leached profiles are the product of the activity of the infiltrating water. In the case of the African surface in Malawi there are three distinct components in the landscape: leached profiles on interfluves; dambo profiles in the bottomlands where there is clay accretion; and dambo-peripheral profiles where, because of incision of the ancient landsurface, former dambo clay is now subject to leaching similar to the interfluves. Leaching of the interfluve profiles, the recharge areas, has resulted in horizon differentiation within the profile such that shallow and deep water are chemically distinctive. Shallow throughflow, with a low EC and low levels of bases and silica, moves laterally downslope towards the dambos, discharging at the seepage zone. Deep water, with a high EC and rich in bases and silica, discharges locally, via by-passes, near the margins of the dambos.

Water levels in the interfluve profiles are anomalously high; some 8 m of what was formerly vadose saprolite is now below the water table. This can be attributed to the replacement of the natural open forest with rain-fed crops. Such an increased head in very low-relief terrain must have resulted in increased vigour of deep water discharge into the dambos and in the maintenance of shallow throughflow towards the dambos throughout the dry season. Since the shallow throughflow passes through that part of the profile where congruent kaolinite dissolution occurs, Al was expected to occur in seepage zone water and in shallow hand-dug wells around the dambos, where the water table is high. The expectation was based not only on Al loss from the interfluve profile but also on the recognition of contemporary alumino-silicate evaporites on the dambo floors. Nevertheless, ICP–OES analyses of filtered and acidified waters failed to find Al within the instrumental detection limit of 0.1 ppm.

In the advanced stages of weathering and leaching in the upper parts of the interfluve profile losses of the elements which would not be expected to be soluble under the ambient conditions, for example Al, Fe, Cr and Ti, clearly indicates that solubility and mobility cannot be regarded as synonymous; loss of these elements involves mobilization in other forms. It follows that the pre-analytical treatment of samples requires further research since clearly in the case of Al the treatment mitigates against its detection.

A substantial part of this research was sponsored by ODA as a component of the British Geological Survey Basement Aquifer Project. Gamma logging was by S. Shedlock. Clay mineral determinations were by D. Morgan, B. Bloodworth, R. L. F. Kay and M. Neal. Logging of the dambo stratigraphy was by E. Freshney. ICP–OES determinations were by D. Miles and J. Cook. Chemical analyses were provided by G. Hendry (Earth Sciences, Birmingham University). Newman and Westhill College, Birmingham provided financial support which enabled D. Bowden to collect shallow well water in Malawi after the Basement Aquifer Project was completed. A British Geomorphological Research Group grant for ICP–OES by BGS is also gratefully acknowledged. I am very grateful to K. Smith (PGIS, Reading University) for help with electron microscopic analysis and to D. Bowden and L. Giusti (School of Geography, Oxford University) for constructive criticism of this paper.

References

ALEVA, G. J. J. 1983. On weathering and denudation of humid tropical interfluves and their triple planation surfaces. *Geol. Mijnbouw*, **62**, 52–73.

ALLEN, D. J. 1987. *Hydrogeological Investigations using Piezometers at Chimimbe and Chikhobwe Dambo sites, Malawi, in August–September 1986.* British Geological Survey Unpublished Report, No. EGARP/WL/87/4.

—— 1988. *Hydrogeological Investigations at Chimimbe Dambo, Malawi, in May–June 1987.* British Geological Survey Unpublished Report, No. WN/88/3.

BELLINGHAM, K. S. & BROMLEY, J. 1973. *The Geology of the Ntchisi-Middle Bua Area*. Malawi Govt. Printer, Zomba, Malawi.

BIRD, M. J., MCFARLANE, M. J. & NEAL, M. 1988. *Porosity and Density Results from Chikhobwe BH. AL1, Malawi*. British Geological Survey Basement Aquifer Project Unpublished Report No. WN/88/9R.

BUDEL, J. 1957. Die 'Doppelten Einebungsflachen' in den feuchten Tropen. *Zeitschrift für Geomorphologie*, N.F. **1**, 201–228.

—— 1977 *Klima-Geomorphologie*, Berlin (Trans. FISCHER, L. & BUSCHE, D. 1982). *Climatic Geomorphology*. Princeton University Press, New Jersey.

FRESHNEY, E. C. 1985. *End of Year Report of the Dambo Stratigraphy Project*. Unpublished British Geological Survey Report.

—— 1987. *Stratigraphy and Origin of Selected Dambos from the Central Region of Malawi*. Unpublished British Geological Survey Report.

GRIM, R. E. 1968. *'Clay Mineralogy'*, 2nd edn, McGraw-Hill, New York.

KAFUNDU, R. D. 1984. *Groundwater Occurrence in the Weathered Basement Complex Rocks of Dowa West, Malawi*. MSc Thesis, University College, London University.

KING, L. C. 1963. 'South African Scenery', 2nd edn, Oliver & Boyd, Edinburgh.

—— 1967. 'Morphology of the Earth', 3rd edn, Oliver & Boyd, Edinburgh.

DE LAPPARENT, U. 1941. Logique des mineraux du granite. *Rev. Scientif.*, 248–292.

LELONG, F. & MILLOT, G. 1966. Sur l'origine des mineraux micaces des alterations lateritiques. Diagenese Regressive-mineraux en transit. *Bull. Ser. Carte geol. Alsace Lorraine* **19**, 271–287.

LISTER, L. A. 1967. Erosion surfaces in Malawi. *Records of the Geological Survey of Malawi* **7** (for 1965), 15–28.

MACKEL, R. 1974. Dambos: a study of morphodynamic activity on the plateau regions of Zambia, *Catena*, **1**, 327–365.

—— 1985. Dambo environments of the Central Plateau region of Zambia, *Zambian Geographical Journal* **35**, 1–17.

MCFARLANE, M. J. 1986. Interpretation of Weathering Profiles, British Geological Survey Basement Aquifer Project, End of Year Report.

—— 1988a. *Some Aspects of the Chemical Characteristics of the Chimimbe Interfluve Core (B3), Malawi*. British Geological Survey Basement Aquifer Project Unpublished Report.

—— 1988b. *S.E.M. of Clay Glaebules in Malawi*. British Geological Survey Basement Aquifer Project Unpublished Report.

—— 1989. Dambos—their characteristics and geomorphological evolution in parts of Malawi and Zimbabwe with particular reference to their role in the hydrogeological regime of areas of African Surface. *In: Groundwater Exploration and Development in Crystalline Basement Aquifers*", **1**, 254–310. Commonwealth Science Publication.

—— & HEYDEMAN, M. T. 1984. Some aspects of kaolinite dissolution by a laterite-indigenous microorganism. *Geographie et Ecologie Tropicale* **8**, 73–91.

——, MILES, D. L. & COOK, J. M. 1988. *Preliminary investigation of water characteristics from a selection of crescent springs in dambos, dambo-peripheral seepage zones and a palaeodambo interfluve profile in Malawi*. British Geological Survey Basement Aquifer Project Unpublished Report.

—— & POLLARD, S. 1989. Some aspects of stone-lines and dissolution fronts associated with regolith and dambo profiles in parts of Malawi and Zimbabwe. *Geographie et Ecologie Tropicale* **11**, 23–35.

——, HENDRY G. & KAY, R. L. F. in prep. Lateritisation associated with the African Surface in central Malawi. *Dambos and Duricrusts Symposium, Second International Geomorphology Conference, Friedrichsdorf, FRG, September 1989*.

SHEDLOCK, S. L. 1990. Borehole geophysical logging within crystalline basement aquifers. *In: Groundwater Exploration and Development in Crystalline Basement Aquifers*, Vol. II, 149–168, Commonwealth Science Council Publication.

WAYLAND, E. J. 1934 Peneplains and some other erosional platforms. *In: Annual Report, Geological Survey of Uganda, 1933*, 77–78.

From WRIGHT, E. P. & BURGESS, W. G. (eds), 1992, *Hydrogeology of Crystalline Basement Aquifers in Africa*
Geological Society Special Publication No 66, pp 131–154.

Geomorphological controls on borehole yields: a statistical study in an area of basement rocks in central Malawi

M. J. McFarlane,[1] P. J. Chilton[2] & M. A. Lewis[2]

[1] *School of Geography, Oxford University, Mansfield Road, Oxford OX1 3TB, UK*
[2] *British Geological Survey, Maclean Building, Wallingford, Oxon OX10 8BB, UK*

Abstract. Data from about 1500 boreholes in an area of African erosion surface in central Malawi were statistically analysed to attempt to correlate terrain characteristics, profiles and yields. The objectives were to seek terrain criteria which (a) allow regional evaluation of hydrogeological potential and (b) facilitate borehole site selection. From the 26 1:50 000 topographic maps analysed, a close map-by-map correlation was shown between extensive areas of dambo (seasonally waterlogged bottomland), low stream frequency and low relief as expressed by mean, median and modal values of the relative relief of individual km squares. Using this map-by-map approach no strong correlations were established between yield, profile and terrain characteristics.

There were only slight variations in mean yield of the two borehole types: those which finished in the regolith and those penetrating the underlying hard rock (55 l/min and 68 l/min respectively). Division of the total borehole data into relative relief subsets showed clear associations of low relative relief, thicker total and thicker saturated regolith and higher yields, where relative relief is <250 ft/km^2. Subsetting the data with respect to the eight minor erosion surfaces which here comprise the African surface also showed significant variations in mean yield associated with these surfaces. Low relative relief emerged as the prime target for borehole siting, with sites at altitudes between the surfaces more likely to be higher yielding than comparably low relative relief sites within the altitudinal range of the minor surfaces. Sites close to the base of inselbergs yielded less well than those in the lowest relief category. The emergence of strong relationships between terrain and borehole performance, regardless of lithological variations, places the effects of rock type on borehole performance of secondary importance to those of geomorphology and leaching history in the study area.

This study allows regional assessment of hydrogeological potential in terms of terrain type. It also provides borehole site selection options within 1 km of a given site, using tabulated mean yields in the relative relief classes in various positions in relation to the minor erosion surfaces. Only the information available on the 1:50 000 topographic maps, i.e. altitude and relative relief as represented by the 50 ft contours, is necessary to apply the siting options.

Malawi has a population of some eight million, 90% of whom live in rural areas. About 35% of the rural population have clean, potable water supplies from boreholes and dug wells with handpumps or by reticulation from protected surface water sources (Smith-Carrington & Chilton 1983). There is little potential for further development of surface water schemes without the construction of costly dams. Thus, provision of improved supplies to the remainder of the dispersed rural population must come from groundwater, and dominantly from basement rocks. There are perhaps 7000 boreholes with handpumps in Malawi and achieving full coverage in a reasonable timescale implies the construction of several hundred more each year for many years to come. There is considered to be scope for improving yields by better siting methods.

Scope of study

Geomorphology plays a dominant role in the occurrence of groundwater in basement areas because this controls the leaching history and hence the nature and depth of the regolith, which provides essential groundwater storage. Thus, as an integral part of the BGS Basement Aquifer Study, a statistical analysis of borehole data was undertaken for part of the Central Region of Malawi, in which a large number of boreholes had been constructed in basement rocks over the last twenty years. The analysis was designed to seek relationships between terrain features, regolith characteristics and borehole performance, for two fundamental purposes.

(a) It should assist in regional assessment of groundwater occurrence. Thus, for example, should there be significant variations in yield in areas of higher or lower relief or on differently aged erosion surfaces, this could allow a degree of prognostication for new drilling programmes in comparable terrain.

(b) It should attempt to provide improved criteria for borehole siting. Given the constraint that a supply must be within reasonable walking distance of the centre it is designed to serve, are there geomorphological options which increase the probability of appropriate borehole performance? Can such options direct the use of, or even obviate the need for costly geophysics?

This last issue is particularly relevant to Malawi. In drier areas on young erosion surfaces, e.g. in southern Zimbabwe, deep weathering is very patchy and geophysics assists the identification of localities where the regolith is capable of providing the necessary storage for hard rock boreholes (Wright 1988). By contrast, in Malawi, dominated by the African erosion surface, deep weathering profiles are extensive and recent drilling has focused on abstraction from shallow holes, completed in the regolith, drilled relatively inexpensively by light-weight rigs. Given this situation, the questions arise: 'Does thicker regolith correlate with higher yields and are these sufficiently higher that it is cost-effective to use geophysics to locate thicker regolith?'

Thus this analysis takes cognizance of serious and pragmatically valid criticism commonly voiced by Third World hydrogeologists, to the effect that statistical analysis is inappropriate to the direct needs of those concerned with the task of deciding where to drill, given a specific centre which requires a water supply. The analysis was undertaken also with the understanding that since those concerned with borehole siting rarely have training in geomorphological techniques, unless geomorphological criteria can be expressed as simple guidelines, they cannot be applied.

Geomorphological and hydrogeological setting

The principal geomorphological components of central Malawi are shown in Fig. 1.

(a) Residuals of ancient (Gondwana and post-Gondwana) surfaces stand as elevated plateaux above the African surface, along the shoulder of the Rift Valley, within which lies Lake Malawi which functions as a local base level for contemporary geomorphological processes.

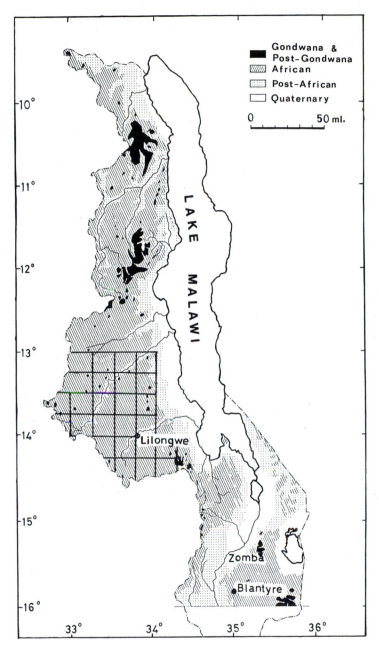

Fig. 1. The location of the study area (26 1:50 000 topographic maps) in relation to the main erosion surfaces, following Lister (1967).

(b) The terrain is dominated by the African erosion surface characteristically expressed as extensive plains of very low relief. Low divides, sometimes with inselbergs, are separated by wide, seasonally waterlogged bottomlands (dambos). The plains decrease in altitude westwards, away from the shoulder of the Rift Valley, a decline attributed to tilting (King 1963, 1967; Lister 1967).

(c) Waxing gradients are associated with incision of the African surface by streams flowing into the Rift Valley.

The study area is dominated by the African surface, characterized by monotonously low relief, becoming more accentuated towards the edge of the Rift Valley. Much of the existing literature on Malawi follows King (1963, 1967) in its assumption that the formation of this land surface is by the mechanism of pediplanation, involving mechanical stripping of surface material, thus implying that direct surface runoff is the main agent of erosion. Recent research (McFarlane 1989b, c) both in Malawi and Zimbabwe, has provided support for an alternative mode of surface formation, etch-planation, which involves differential leaching to the extent that the weathered rock (saprolite) collapses, leaving a thick surficial residuum. This implies that planation is achieved not by the effects of direct runoff but by infiltrating water which either becomes shallow throughflow or recharges the main groundwater system and this implication receives substantial independent support from hydrological research in Zimbabwe (Bullock 1988, in prep.). The seasonally waterlogged bottomlands (dambos) which are streamless where the African surface is most subdued, are deduced to occupy sites where the lithology is most susceptible to differential leaching, saprolite collapse and surface lowering (McFarlane 1989a; McFarlane & Lewis in prep.).

Methods of study

The study area, approximately $20\,000\,km^2$ of the central plateau area of Malawi (Fig. 1), is underlain by metamorphic rocks of the Precambrian to Lower Palaeozoic Basement Complex. The study was carried out in two parts. In the first phase, possible relationships were examined on a map-by-map basis for the 26 1:50 000 topographic map sheets comprising the study area. In the second phase all boreholes were treated in the context of their immediate vicinity.

Phase 1

1. Borehole data

Information obtained at the time of construction was extracted from the national borehole archive held by the Water Department of the Ministry of Works in Lilongwe. In total, records from about 1500 boreholes were entered into a commercially available database system to permit rapid statistical analysis. The information entered included construction details of the boreholes, water levels, yields and geological descriptions of the material penetrated by the drilling, provided either by the driller or by a supervising field geologist. This often limited lithological information was interpreted by project staff to provide estimates of the thickness of the regolith at each site. Thus, regolith thickness as defined here is the vertical distance from the ground surface to the basal surface of weathering. Similarly, saturated regolith thickness is the depth from the rest water level to the basal surface of weathering. Specific capacity, yield in litres per minute per metre of drawdown, would be the best measure of borehole productivity. However, this information was available for relatively few (27%) of the boreholes, and yield was therefore used.

Boreholes were also divided into two subsets: regolith holes which drew entirely from weathered material, and holes which penetrated the underlying hard rock. Some of the boreholes had incomplete information, i.e. water levels but not lithological descriptions or vice versa, which explains the differences in borehole totals from Table 4 onwards.

2. Terrain characterization

2.1. Relief

Characterization of the terrain in areas of such subdued relief presents difficulty. Earlier studies which attempted to relate streamflow parameters to terrain characteristics in a selection of catchments in Malawi and Zimbabwe (Drayton *et al.* 1980; Meigh 1987) followed the generalization that slopes in a catchment are directly related to the gradient of the main stream (Benson 1959; Strahler 1950) and in these catchment studies, relief was represented by mainstream gradient. While this generalization may apply in monocyclic situations, it is inappropriate in polycyclic tropical areas, where streams may pass, with equal lack of gradient expression, between interfluves which are very flat, or which bear a flight of minor low relief surfaces, or which are crowned by inselberg-like residuals standing above the generally flat surface or surfaces. Detailed cartographic analysis of a selected area within the main study area allowed the deduction that the African Surface here comprises a flight of eight minor surfaces, apparently undeformed, descending westwards away from the Rift shoulder (McFarlane 1989c). Thus, there are two essential components to terrain characterization: relative relief (regardless of its assignment to a specific surface) and its assignment to the appropriate surface or situations between surfaces. Since, on each of the map sheets, surviving areas of the minor surfaces occur as complicated mosaics, it was impracticable to include ages of minor surfaces as a terrain-characterization component of this map-by-map phase of the study. Only relative relief, independent of surface age, was considered. Nevertheless, the awareness of polycyclicity and the very obvious presence of residuals on some interfluves contributed to the conclusion that mainstream gradient is likely to be an inappropriate expression of the relief of the terrain. This conclusion received support from the lack of correlation between catchment relief, thus expressed, and baseflow, while a small negative correlation was found between dambo area and baseflow (Meigh 1987). Since, from observation of topographic maps, it is very clear that the lower the relief the greater is the area of dambo, that is, dambo area correlates negatively with relief, it was surprising that a baseflow/dambo area relationship was expressed but a baseflow/relief relationship was unidentifed.

In short, a problem with the catchment studies appeared to lie in inappropriate characterization of relief. Hence a different approach was used in this study. The relief of the individual map sheets was derived from the relative relief of individual kilometre squares by counting the 50 ft contours crossing it and subtracting one (NC-1), to give the relative relief or range of altitude in the squares (broadly representative of the gradient within it). Higher numbers thus indicate greater relative relief and the data was expressed as relief classes, as shown in Table 1. The relief on the map sheets was expressed as the mean, median and modal values of the square kilometre values for each sheet and % area $< 50\,\text{ft/km}^2$ and $\% < 150\,\text{ft/km}^2$ were calculated.

Table 1. *Relative relief classes. Representation of the relative relief of each km² was made by subtracting one from the number of different 50ft interval contours crossing it (NC-1)*

NC-1	Relative relief (ft/km²)	Class
0	<50	A
1, 2	50–150	B
3, 4	150–250	C
5–9	250–500	D
10–13	500–700	E
14+	>700	F

2.2. Dambo area

The area of dambo, as shown on the 1:50 000 topographic maps was checked against air photo representation and in selected areas in the field. Although numerous tiny, discrete patches of dambo were identified on the air photos and their existence confirmed in the field (McFarlane 1986), the main areas of dambo were found to be represented sufficiently well on the topographic maps for the purposes of this study. Their total area was determined by cutting them out of a dye-line copy, weighing dambo and non-dambo cuttings and thus expressing dambo area as percentage of the total area. This method is quicker and less conducive to error than the use of planimeter measurements. The total dambo area was similarly subdivided into streamless dambo and dambo with streams following air photo and field confirmation of adequate representation of the streams on the topographic maps.

2.3. Stream frequency

In the case of the catchment studies (Drayton *et al.* 1980; Meigh 1987) the assumptions were made that channels occupy central positions in all dambos and that discrete areas of dambo are linked to the main dambo systems by streams, although these postulated channels are not represented on the 1:50 000 topographic maps of Malawi. Fieldwork showed, however, that these assumptions are incorrect and that the streams depicted on the maps are an appropriate representation of the existence of channels. Thus, in this study, stream frequency was derived from streams shown on each map and was expressed as the number of junctions/km², using only the mapped streams.

Results

Phase 1

Correlations of terrain parameters are shown in Table 2. Negative correlations were found between relative relief, and the total and streamless dambo areas. A high positive correlation ($r = 0.83$) was found between both total and streamless dambo area and percent area with relative relief less than 50ft/km², and a very high positive correlation ($r = 0.93$) between median relative relief value and stream frequency. This confirmation of the observed field relationships, that is, the lower the relief the greater the dambo area and the less frequent the streams, lends support to the appropriateness of the techniques used in this study to characterize the relief of the terrain, with the median relative relief value emerging as the most appropriate parameter for relief characterization in a regional approach.

Table 2. *Pearson's correlation coefficients of terrain parameters on a map-by-map basis*

	1 TDA	2 SDA	3 Mode	4 Median	5 Mean	6 % <0	7 % <1
SDA	0.94						
Mode	−0.58	−0.54					
Median	−0.74	−0.69	0.87				
Mean	−0.71	−0.68	0.78	0.94			
% <0	0.83	0.83	−0.59	−0.80	−0.83		
% <1	0.78	0.73	−0.68	−0.89	−0.95	0.89	
SJK	−0.73	−0.69	0.80	0.93	0.91	−0.77	−0.90

1 (TDA): total dambo area.
2 (SDA): streamless dambo area.
3, 4, 5 (Mode, Median, Mean): modal, median and mean values of relative relief of individual kilometre squares.
6, 7 (% <0, % <1): percentage of map sheet with kilometre squares having relative relief <50 ft/km^2 and <150 ft/km^2.
SJK: stream junctions/km^2.
($r \geqslant 0.45$ significant at 1% level)

Correlations of regolith thickness, saturated regolith thickness and yield with terrain parameters are shown in Table 3. There is a weak, positive correlation of regolith thickness with total dambo area and low relative relief and stronger correlations between saturated regolith thickness and dambo area, and between saturated regolith thickness and median relative relief. There are only very weak correlations between yield and low relative relief. Thus this approach has provided relatively little useful information about borehole prospects. However, since it provides reasonable support for the observation that water levels are generally shallower where relief is low, an obvious relationship, then the poverty of other relationships is unlikely to be attributable to poor data quality. This raises the following possibilities:

(a) there are no relationships between borehole prospects, terrain and weathering profile characteristics;
(b) there are subsets within the main data set, showing relationships which are obscured by this bulk treatment of the data;
(c) lithological variations exert a stronger control on borehole performance than do terrain and weathering profiles.

Phase 2

Of the options identified in Phase 1 of the analysis, only the existence or otherwise of relationships between yield, terrain and weathering profiles could be further examined, as the data on the nature of the hard rock was too sketchy for appropriate classification and was very poor or absent in the case of holes finishing in the regolith.

Table 3. *Pearson's correlation coefficients of regolith thickness, saturated regolith thickness, total saturated aquifer and yield with terrain parameters, on a map-by-map basis (for definition of parameters see Table 2).*

	Yield	Reg. thick.	Sat. reg. thick.	Tot. sat. aqu. thick.	Log yield	Log reg. thick.	Log sat. thick.	Log tot. sat. aqu. thick.
TDA	0.40	0.48	0.67	0.77	0.42	0.47	0.65	0.78
SDA	0.33	0.37	0.53	0.82	0.36	0.37	0.51	0.79
mode	−0.46	−0.51	−0.51	−0.45	−0.49	−0.56	−0.55	−0.48
median	−0.58	−0.48	−0.62	−0.57	−0.61	−0.53	−0.65	−0.60
mean	−0.46	−0.45	−0.57	−0.55	−0.50	−0.49	−0.60	−0.57
% <0	0.51	0.35	0.51	0.79	0.52	0.36	0.51	0.77
% <1	0.44	0.46	0.61	0.60	0.48	0.48	0.62	0.61
SJK	−0.59	−0.37	−0.55	−0.48	−0.64	−0.41	−0.57	−0.49
log TDA	0.52	0.53	0.65	0.56	0.56	0.58	0.69	0.58
log SDA	0.52	0.45	0.60	0.56	0.56	0.49	0.62	0.57
log mode	−0.44	−0.43	−0.50	−0.28	−0.47	−0.46	−0.48	−0.27
log median	−0.48	−0.47	−0.70	−0.50	−0.54	−0.50	−0.68	−0.50
log mean	−0.42	−0.34	−0.48	−0.71	−0.45	−0.36	−0.49	−0.69
log % <0	0.50	0.16	0.43	0.55	0.54	0.15	0.41	0.55
log % <1	0.48	0.51	0.59	0.49	0.52	0.55	0.63	0.52
log SJK	−0.56	−0.32	−0.48	−0.67	−0.57	−0.35	−0.50	−0.64

($r \geqslant 0.45$ significant at 1% level)

Individual holes were assessed in terms of two terrain parameters:

(a) the relative relief of the kilometre square within which the borehole was located, classified as before;

(b) the position of the borehole in relation to the eight minor surfaces identified as components of the African surface.

Since the altitudinal separation of these minor surfaces ranges from 35 m to 70 m, which is greater than the mean depth of weathering (*c.* 29 m) the possibility exists that individual surfaces may bear profiles which are significantly different in terms of hydrogeological potential. Because, in this study area, the minor surfaces appear to be undeformed, boreholes could be assigned to positions, either within the range of a particular surface or falling between surfaces, purely on the basis of their altitude.

The boreholes used in this analysis were of two types, those which finish in the regolith and those which penetrate hard rock below the regolith. Differences in performance of these two groups were examined. The data shown in Fig. 2 can be interpreted to imply that boreholes completed in the regolith may sometimes miss additional contributions to yield from deeper fractures. However, the differences are relatively small and hence the use of both types of borehole for this analysis was not considered to be a likely reason for poor correlations between performance and terrain parameters. Indeed, the data shown in Table 4 strongly suggest that the mean yield benefits of intercepting quartz stringers (most commonly observed as features within the regolith) are as great as intercepting fractures in the bedrock, which lends further support to the contention that it is valid to group regolith holes and deeper holes in an attempt to seek correlations between yield and terrain parameters.

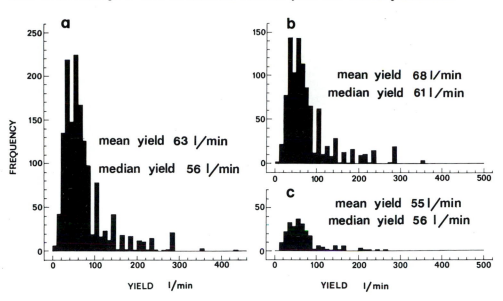

Fig. 2. Frequency distributions of borehole yields. (a) Total holes. (b). Holes penetrating hard rock. (c) Holes finishing in regolith.

Table 4. *Mean yields of boreholes terminating in regolith and in bedrock and mean yield variations of holes with and without record of quartz stringers (predominantly observed in the weathering profile) and fractures in the bedrock*

Boreholes	Mean yield (l/min)
Total (1462)	63
Terminating in regolith (256)	55
Terminating in bedrock (968)	68
With quartz stringers recorded (258)	82
With no quartz stringers recorded (1204)	59
With fractures recorded (397)	80
With no fractures recorded (1065)	57

Scatter plots of yield against regolith thickness and against saturated regolith thickness for all boreholes, regolith boreholes and regolith + rock boreholes indicated very poor correlations for these large data sets (Fig. 3) which ignored relative relief and position in relation to erosion surfaces.

Breakdown of the mean yield data into relative relief subsets, located on or between erosion surfaces, are given in Table 5. The mean yield of all holes decreases through relative relief categories A to C, from 72.6 l/min to 48.9 l/min, holes in these

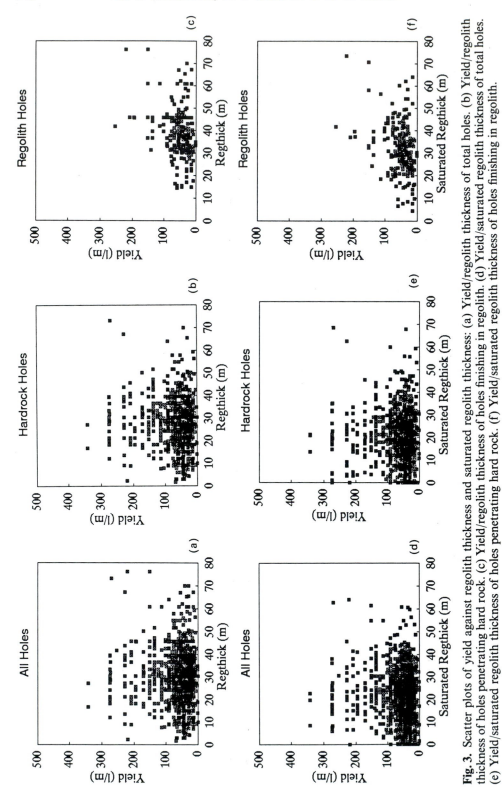

Fig. 3. Scatter plots of yield against regolith thickness and saturated regolith thickness: (a) Yield/regolith thickness of holes penetrating hard rock. (b) Yield/regolith thickness of holes finishing in regolith. (c) Yield/regolith thickness of total holes. (d) Yield/saturated regolith thickness of total holes. (e) Yield/saturated regolith thickness of holes penetrating hard rock. (f) Yield/saturated regolith thickness of holes finishing in regolith.

classes comprising 95% of the total. Class D does not continue this trend and classes E to F have too little data to be meaningful. Thus, in areas with relative relief of less than 250 ft/km^2, higher yields are associated with progressively flatter land. This same relationship is maintained with remarkable uniformity when the data set is split into boreholes on the erosion surfaces and between erosion surfaces, implying that flat land is more important for high borehole yield than position on or between erosion surfaces.

Table 5. *Mean yields (total and in relative relief subsets) on and between erosion surfaces*

Relative relief	Mean yield (l/min)	Mean yield on an erosion surface (l/min)	Mean yield between erosion surfaces (l/min)
Total	63.1 (1447)	64.0 (635)	62.4 (812)
Class A (35.7%)	72.6 (518)	73.6 (234)	71.7 (284)
Class B (53.5%)	58.4 (774)	58.8 (334)	58.2 (440)
Class C (6.0%)	48.9 (87)	52.0 (37)	46.6 (50)
Class D (3.9%)	62.2 (56)	61.0 (25)	63.2 (31)
Class E (0.4%)	60.5 (6)	45.0 (1)	63.6 (5)
Class F (0.4%)	61.8 (6)	66.5 (4)	52.5 (2)

Within each relative relief class, boreholes were classified according to position in relation to specific erosion surfaces, as shown in Table 6. Mean yields of total holes on the erosion surfaces (E1, E2, etc.) varied considerably, alternating high and low (E1 = 72.0, E2 = 57.4, E3 = 64.9, E4 = 53.9, E5 = 68.0) except for the lowest three which showed increasing yield with reduced altitude (E6 = 56.6, E7 = 65.9, E8 = 73.9). Similar patterns were followed in relative relief classes A and B, data for the other classes being too few. The range of mean values associated with the different surfaces (73.9 to 53.9) is not far short of the range in the relative relief classes, A to C (72.6 to 48.9), suggesting that the subdivision of the African surface into minor surfaces, for hydrogeological purposes, appears to be justified in terms of their profiles having different hydrogeological characteristics.

Earlier work had indicated an association of more aggressive leaching and more permeable regolith with breaks of slope, in African Surface situations (Wright *et al.* 1988; McFarlane 1989*b*). This raised the possibility that the higher yielding profiles may lie not on the minor erosion surfaces but between them. The first three columns of Table 6 show the data relevant to this question. From column 1, comparing the yield on each surface with the yield on the 'break' below it, there is no sign of consistently higher yields in the 'breaks'. However, in relief class A (column 2), the yields in the breaks below the erosion surfaces are consistently higher than the equivalent mean value in column 1 (e.g. 93.3 compared with 59.5 for 4050 ft +, 71.5 compared with 66.5 for 3950 ft +, etc.) and in four of the seven class A cases the

Table 6. *Borehole yields (total and in relative relief subsets) on and between specific erosion surfaces, designated E1, E2 etc., in order of descending altitude*

Altitude range (ft)	Total Mean (l/min)	Class A Mean (l/min)	Class B Mean (l/min)	Class C Mean (l/min)	Class D Mean (l/min)	Class E Mean (l/min)	Class F Mean (l/min)
4500 – 5625	60.1 (16)	27.0 (1)	58.3 (3)	52.0 (6)	74.7 (6)	—	—
(E1) 4450 – 4500	72.0 (13)	—	77.9 (8)	62.6 (5)	—	—	—
4300 – 4450	58.1 (30)	34.0 (1)	66.9 (15)	43.4 (7)	63.0 (5)	42.5 (2)	—
(E2) 4250 – 4300	57.4 (14)	171.0 (1)	43.7 (7)	63.8 (4)	35.5 (2)	—	—
4150 – 4250	53.5 (46)	58.6 (7)	58.3 (29)	26.5 (6)	47.5 (2)	—	52.5 (2)
(E3) 4100 – 4150	64.9 (60)	78.6 (10)	68.0 (31)	39.5 (8)	63.1 (8)	—	58.3 (3)
4050 – 4100	59.5 (44)	93.3 (11)	48.2 (25)	48.4 (5)	46.0 (2)	52.0 (1)	—
(E4) 4000 – 4050	53.9 (90)	64.4 (26)	48.8 (51)	30.0 (4)	61.7 (7)	45.0 (1)	91.0 (1)
3950 – 4000	66.5 (57)	71.5 (19)	66.3 (35)	36.0 (3)	—	—	—
(E5) 3900 – 3950	68.0 (84)	77.9 (25)	65.0 (53)	39.3 (4)	79.0 (2)	—	—
3800' – 3900	70.3 (175)	73.7 (65)	69.5 (97)	49.8 (5)	58.5 (6)	90.5 (2)	—
(E6) 3750 – 3800	56.6 (119)	54.2 (41)	54.8 (69)	74.9 (7)	102.5 (2)	—	—
3650 – 3750	55.3 (239)	59.1 (82)	53.1 (147)	48.4 (9)	114.0 (1)	—	—
(E7) 3600 – 3650	65.9 (141)	73.4 (74)	59.0 (60)	55.7 (3)	38.3 (4)	—	—
3550 – 3600	68.3 (105)	79.6 (54)	56.1 (50)	—	68.0 (1)	—	—
(E8) 3500 – 3550	73.9 (114)	87.7 (57)	61.0 (55)	36.0 (2)	—	—	—
2940 – 3500	64.1 (100)	81.1 (44)	47.4 (39)	57.9 (9)	59.4 (8)	—	—
		(35.7%)	(53.5%)	6.0%	(3.9%)	(0.4%)	(0.4%)

mean yield in the break is higher than that on the erosion surface above. This appears to lend some support to the hypothesis that break zones are favourably high yielding, but only where the relief is very low. In practical terms it means that there is a better chance of obtaining higher borehole yields on flat land below the erosion surface than on flat land on the surface itself. In the case of relative relief class B no such generalization emerges. This situation is summarized in Fig. 4, which illustrates the yield benefits of selecting class A sites on the erosion surface (horizontal arrows) and between them, as well as the benefits of selecting class A sites at altitudes below the surfaces (vertical, circled arrows).

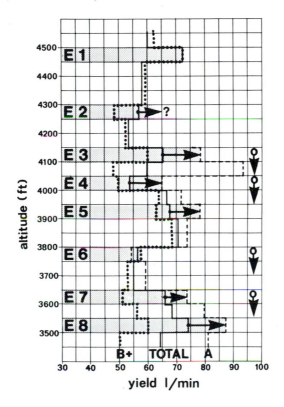

Fig. 4. The yield benefits of selecting sites in relative relief class A, on the erosion surfaces (horizontal arrows) and between them, and the additional benefits of selecting class A sites at altitudes below the surface (vertical, circled arrows).

Table 7 shows data for regolith thickness (as for borehole yield in Table 5). There appears to be little difference between mean regolith thicknesses on and between erosion surfaces in three of the four lowest relief categories. Mean regolith thickness decreases progressively through relative relief classes A to E (from 31.6 to 13.5 m), these comprising over 99% of the boreholes. Similar decreases occur on erosion surfaces from class A to class C (95% of holes) and also between the erosion surfaces. Unlike yield (Table 5), the pattern is continued into the higher relief categories. In conjunction, Tables 5 and 7 indicate a positive correlation between regolith thickness and yield, both decreasing as relative relief increases in relative relief classes A to C which comprise all but 5% of the holes. Since relative relief correlates with dambo area and stream frequency (Phase 1) the data in Table 7 may be further interpreted as establishing a relationship between high yields, thick regolith, large dambo area and low stream frequency.

Table 7. *Thickness of regolith (total and in relative relief subsets) on and between erosion surfaces*

Relative relief	Mean regolith thickness (m)		Mean regolith thickness on an erosion surface (m)		Mean regolith thickness between erosion surfaces (m)	
Total	29.4	(1261)	29.0	(549)	29.7	(712)
Class A 38.1%	31.6	(480)	31.3	(214)	31.9	(266)
Class B 51.6%	29.1	(651)	29.1	(280)	29.1	(371)
Class C 5.7%	24.4	(72)	20.1	(30)	27.4	(42)
Class D 3.9%	22.5	(49)	21.9	(22)	22.9	(27)
Class E 0.6%	13.5	(7)	9.6	(3)	16.4	(4)
Class F 0.2%	14.1	(2)	—		14.1	(2)

Table 8 shows data for saturated regolith thickness which decreases consistently as relief increases in the case of all boreholes and also of boreholes on and between erosion surfaces. As for regolith thickness, the decrease of saturated regolith thickness with increased relief can be extended to a correlation with decreasing yield and in terms of terrain features it makes the link between low yield, few dambos and high stream frequency. In geographical terms, these data indicate that in a westward direction, away from the Rift shoulder towards the Bua river, where the younger, lower and more extensively preserved surfaces occur (McFarlane, 1989c), yields can be expected to be significantly higher than near the Rift Valley shoulder where relative relief increases. This is consistent with the low yields of two recent borehole drilling projects, in Dowa West and Lilongwe NE, (with mean yields of 36 l/min and 54 l/min respectively), which are located in relatively high relief terrain in the east of this study area.

Table 9 shows regolith thicknesses associated with relief classes on and between specific minor surfaces. As concerns total boreholes as well as holes both on and between erosion surfaces, regolith is thickest on the youngest surfaces, decreasing to a minimum at 4100 ft to 4300 ft and increasing again at higher altitudes. In relief class A, adequate data are only available for the five lowest surfaces. As with total boreholes, these show generally increasing regolith thickness on lower surfaces. In higher relief classes this pattern does not emerge.

Table 8. *Thickness of saturated regolith (total and in relative relief subsets) on and between erosion surfaces*

Relative relief	Mean saturated regolith thickness (m)		Mean saturated regolith thickness on an erosion surface (m)		Mean saturated regolith thickness between erosion surfaces (m)	
Total	21.5	(1243)	21.1	(537)	21.8	(706)
Class A 38.3%	24.3	(476)	24.0	(212)	24.6	(264)
Class B 51.8%	20.5	(644)	20.4	(276)	20.6	(368)
Class C 5.4%	16.2	(67)	12.1	(25)	18.6	(42)
Class D 3.9%	16.0	(48)	14.4	(22)	17.5	(26)
Class E 0.3%	11.9	(4)	—	—	11.9	(4)
Class F 0.3%	4.4	(4)	1.7	(2)	7.0	(2)

Table 10 shows saturated regolith thicknesses pertaining to relief classes on and between specific erosion surfaces. Patterns are generally similar to those of regolith thickness shown in Table 9, but both in the case of total boreholes and those in class A, the numerical trends are less pronounced.

Although in all of these parametric inter-relationships considerable consistency can be found in the lower relief classes, in the higher relief categories, classes D to F, similar trends of relationships are not consistently continued. Although the data points are relatively few, they warrant further attention because, in this African surface context, the apparently high relief situation of these boreholes is largely attributable to their location near the foot of inselbergs, such that the berg clips the edge of the kilometre square (Fig. 5). Such situations have long been regarded as of particular interest to hydrogeologists since there is evidence to the effect that precipitation shed from the berg facilitates deep weathering around the base (Ruxton 1958; Thomas 1974; McFarlane 1989c). However, boreholes in such situations also tend to suffer form a very high range of water-table oscillation so that the profile may have a thin dry-season saturated zone or may dry out completely in cases where the berg is smooth, poorly fractured and incapable of storage. If relief classes D to F are grouped, the mean regolith thickness is low, 21.1 m. Mean saturated thickness is also low, at 14.9 m. Nevertheless, mean yield is relatively high, 62.0 l/min, lying between class A and class B values. The high yield despite low water levels would appear to reflect superior saprolite permeability in the lower parts of the weathering profile in such aggressively leached situations. Nevertheless, the mean yield in boreholes closest to the bergs is lower than the mean yield for all boreholes and this suggests that in close-to-the-berg situations, the optimum site would not be close to the berg, but on flat land at least 1 km distant from the base.

Table 9. *Thickness of regolith (total and in relative relief subsets) on and between specific erosion surfaces, designated E1, E2 etc., in order of descending altitude.*

Altitude range (ft)	Total Mean (m)	Class A Mean (m)	Class B Mean (m)	Class C Mean (m)	Class D Mean (m)	Class E Mean (m)	Class F Mean (m)
4500 – 5625	32.2 (14)	31.1 (1)	45.8 (2)	32.6 (5)	27.4 (6)	—	—
(E1) 4450 – 4500	30.4 (12)	—	36.0 (8)	19.0 (4)	—	—	—
4300 – 4450	26.8 (27)	14.9 (1)	29.1 (15)	33.4 (5)	18.4 (4)	16.0 (2)	—
(E2) 4250 – 4300	23.6 (13)	27.5 (1)	24.3 (6)	22.2 (4)	22.3 (2)	—	—
4150 – 4250	24.8 (28)	19.8 (2)	27.1 (17)	26.4 (5)	16.3 (2)	—	14.1 (2)
(E3) 4100 – 4150	23.8 (49)	25.6 (8)	27.5 (27)	15.6 (6)	17.8 (6)	9.6 (2)	—
4050 – 4100	26.0 (34)	29.3 (10)	26.3 (18)	21.7 (4)	15.9 (2)	—	—
(E4) 4000 – 4050	26.1 (72)	30.9 (22)	26.4 (39)	16.0 (3)	16.4 (7)	9.6 (1)	—
3950 – 4000	29.2 (52)	28.4 (17)	30.1 (32)	23.0 (3)	—	—	—
(E5) 3900 – 3950	26.1 (70)	29.6 (21)	24.9 (44)	20.5 (4)	29.0 (1)	16.7 (2)	—
3800 – 3900	29.2 (161)	31.9 (68)	27.7 (82)	33.0 (4)	21.5 (5)	—	—
(E6) 3750 – 3800	31.1 (116)	31.0 (42)	31.6 (67)	26.4 (5)	28.1 (2)	—	—
3650 – 3750	30.5 (223)	31.5 (81)	30.3 (132)	24.4 (9)	20.1 (1)	—	—
(E7) 3600 – 3650	30.4 (124)	31.5 (67)	29.0 (51)	27.5 (2)	32.8 (4)	—	—
3550 – 3600	31.4 (94)	32.1 (50)	30.4 (43)—	38.1 (1)	—	—	—
(E8) 3500 – 3550	32.5 (93)	33.0 (53)	32.7 (38)	14.2 (2)	—	—	—
2940 – 3500	30.6 (79)	36.1 (36)	26.2 (30)	26.2 (7)	25.1 (6)	—	—
		(38.1%)	(51.6%)	(5.7%)	(3.9%)	(0.6%)	(0.2%)

Table 10. *Thickness of saturated regolith (total and in relative relief subsets) on and between specific erosion surfaces, designated E1, E2 etc., in order of descending altitude*

Altitude range (ft)	Total Mean (m)	Class A Mean (m)	Class B Mean (m)	Class C Mean (m)	Class D Mean (m)	Class E Mean (m)	Class F Mean (m)
4500 – 5625	21.4 (14)	22.0 (1)	27.9 (2)	20.4 (5)	20.0 (6)	—	—
(E1) 4450 – 4500	20.2 (11)	—	23.9 (8)	10.3 (3)	—	—	—
4300 – 4450	18.2 (27)	5.8 (1)	18.1 (15)	27.5 (5)	14.6 (4)	9.1 (2)	—
(E2) 4250 – 4300	15.9 (13)	24.7 (1)	16.3 (6)	14.9 (4)	12.3 (2)	—	—
4150 – 4250	18.8 (27)	16.5 (2)	21.9 (16)	16.3 (5)	14.7 (2)	—	7.0 (2)
(E3) 4100 – 4150	17.4 (47)	20.1 (7)	21.3 (27)	6.1 (5)	11.5 (6)	—	1.7 (2)
4050 – 4100	19.2 (34)	21.8 (10)	20.5 (18)	12.0 (4)	9.4 (2)	—	—
(E4) 4000 – 4050	19.9 (71)	24.3 (22)	19.1 (39)	14.7 (3)	12.7 (7)	—	—
3950 – 4000	21.8 (52)	21.3 (17)	22.9 (32)	13.3 (3)	—	—	—
(E5) 3900 – 3950	19.3 (68)	23.5 (21)	17.7 (43)	15.6 (3)	10.7 (1)	—	—
3800 – 3900	22.3 (157)	25.3 (66)	20.1 (81)	25.0 (4)	16.2 (4)	14.6 (2)	—
(E6) 3750 – 3800	22.3 (111)	23.6 (42)	21.8 (64)	15.8 (3)	22.3 (2)	—	—
3650 – 3750	21.8 (222)	24.3 (81)	20.8 (131)	15.4 (9)	17.1 (1)	—	—
(E7) 3600 – 3650	21.6 (123)	23.4 (66)	19.8 (51)	12.1 (2)	19.7 (4)	—	—
3550 – 3600	23.4 (94)	24.8 (50)	21.6 (43)	—	32.0 (1)	—	—
(E8) 3500 – 3550	24.0 (93)	25.5 (53)	22.7 (38)	9.1 (2)	—	—	—
2940 – 3500	22.1 (79)	27.2 (36)	17.5 (30)	19.2 (7)	18.9 (6)	—	—
		(38.3%)	(51.8%)	(5.4%)	(3.9%)	(0.3%)	(0.3%)

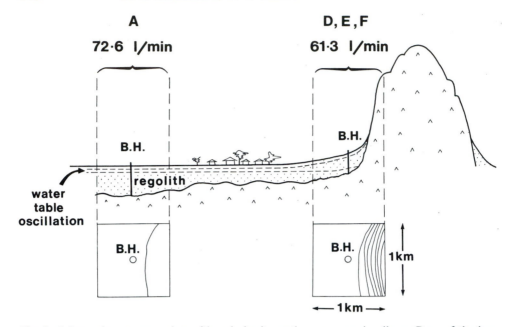

Fig. 5. Schematic representation of borehole site options near an inselberg. Base-of-the-berg sites, classed as 'high relief' (D, E and F), have relatively high yields despite thin regolith. Nevertheless yields are lower than for sites in the lowest relative relief class (A), more distant from the berg.

Discussion

There was originally some doubt about whether the existing borehole data were adequate for such a study since there was no systematic observational framework designed to be of specific relevance to such statistical analysis. Treatment of the entire data sets on a map-by-map basis provides close correlations between low relief, extensive areas of dambo and low stream frequency and this is consistent with other publications which observe a reduction of dambo area as slopes increase (Turner 1985). Although this treatment failed to provide good correlations between borehole performance and terrain features, an essential target of this research, nevertheless it showed that the methods herein selected for characterization of the relief in digital form are appropriate. It also indicated that the borehole data are adequate, and worthy of further enquiry, by the statistically demonstrated relationship between saturated regolith thickness and low relief. Had the borehole data failed to confirm this self-evident relationship, it could have been justifiably regarded as of highly suspect quality.

The map-by-map analysis having established this situation, it became appropriate to consider the poverty of correlation of borehole performance and terrain characteristics in terms of the occurrence of conflicting 'signals' from the geomorphological sub-sets, relative relief and the minor erosion surfaces within the 'broad-brush' African surface. Sufficiently detailed information on geology, from the borehole data, was lacking for an examination of this influence on performance. However, of the two criteria for site selection, geology and geomorphology, the latter is more relevant in terms of accessibility for use in Malawi. It is already known, for example, that biotite-rich facies of the gneissic metasediments are more susceptible to weathering and they weather more deeply than the more basic or quartzo-feldspathic facies (Eggler *et al.* 1969; Thomas 1974; McFarlane 1989*b*). However, it is not known if the advantages of deeper weathering outweigh the disadvantages of the lower permeability of the more clayey saprolite which is the product of weathering of such rocks. In short, we still do not know which rock types should be targeted, to hydrogeological advantage. With detailed information on lithology from the boreholes, it might have been possible to place rock types in a preferential hierarchy. However, even with this knowledge it would be difficult to take advantage of it because in Malawi the regolith on the African surface is sufficiently thick to obscure much of the lithology. Thus, the reality of the Malawian situation is that the accessible guidelines for borehole siting are predominantly topographic. Even where geophysical methods are used to assist borehole siting, the initial selection of an area for geophysical appraisal can do no better than be based on terrain features. The cost-effectiveness of geophysical surveys in such a context is debatable, particularly with the growing emphasis on relatively cheaply and rapidly drilled holes finishing in the regolith. Thus, with or without a geophysical survey, the use of topographic criteria for site selection is of fundamental importance. In African surface areas, where the relief is so subdued that topographic variations are often difficult to identify, let alone evaluate in terms of hydrogeology, it is hardly surprising that social factors usually dominate in initial site selection. In short, geological guidelines are effectively inaccessible here and topographic guidelines offer the best route forward towards improved borehole siting.

The study of individual borehole sites linked high yields with low relief and thick saturated regolith, this in turn establishing a relationship between high yields, low stream frequency and large areas of dambo. In effect this confirms the hypothesis that dambos are essentially contemporary groundwater discharge sites rather than 'fossil' stream systems and goes some way towards explaining other observations concerning the distribution of dambos. For example, the geomorphological situation of the African surface in Zimbabwe is quite different from Malawi, the surface surviving as an elevated strip along the main watershed between drainage to the northwest and southeastwards to the Indian Ocean. There, some extensive areas of relatively low relief have conspicuously few dambos. This may now be interpreted in terms of low water tables, following encroachment by the incising streams of the younger erosion surfaces. Similarly, observations to the effect that dambos only occur where the precipitation is above a certain threshold (Pedro 1968; Thompson 1969; Purves 1976) may also reflect low water tables as a result of low recharge. Thus, climate and geomorphology combine to control the extent of dambo area and in both Malawi and Zimbabwe there is evidence of reduced dambo area during drier periods in the recent past (McFarlane & Whitlow in press).

The association of higher yields with thicker saturated zones may support the belief that permeability of saprolite below the water table increases with depth, as was indicated by the results of piezometric studies in selected profiles (Allen 1987; McFarlane 1989*b*). Although comparison of holes which finish in regolith and those which penetrate more deeply into the hard rock showed greater mean yields in the case of deeper holes (55 l/min as compared with 68 l/min), the difference is relatively slight. The deeper holes have the advantage of intercepting fractures in fresh rock as well as incipient fractures which have been opened by weathering in the saprock, where the early stages of weathering penetration of the fresh rock occur. This apparently small advantage resulting from fracture interception by deeper holes may appear surprising, but is consistent with the suggestion that fractures stay open in the saturated saprolite and that quartz stringers function as efficient by-passes (McFarlane 1989*b*). Thus, these data give support to the contention that it may not be cost-effective, in terms of increased yield, to drill the more expensive holes which penetrate hard rock, rather than holes which finish in the regolith, except in special circumstances, and clearly the thicker the saturated saprolite the greater the chances of intercepting stringers or open fractures, this outweighing the influence of variation in primary permeability associated with different lithologies.

It is intriguing that total thickness of saturated aquifer correlates strongly with low relief (Table 3). Total saturated aquifer thickness clearly depends on the total drilled depth, which might be expected to be random in the sense that drilling stopped when the rate was unacceptably low. This correlation means that holes were consistently drilled more deeply into bedrock where the land was flat and regolith and saturated regolith thick. Confirmation of this deduced bias in drilling comes from close correlations between total borehole depth and total dambo area ($r = 0.68$), streamless dambo area ($r = 0.73$) and percent of land with relative relief less than 50 ft/km^2 ($r = 0.70$) and with less than 150 ft/km^2 ($r = 0.53$). The apparent illogicality of this bias in drilling can be explained in terms of the drillers' criteria, maximum depth penetration but within an appropriate overall drilling rate. Thus, the thicker the easily drilled regolithic component of the profile, the deeper it is acceptable to penetrate the slowly drilled hard rock and still have an appropriate overall drilling rate. Having established that higher yields are associated with flat land, thick regolith and thick saturated regolith, because holes penetrated hard rock more deeply in such situations this raises the possibility that the higher yields may result from this deeper penetration rather than from the occurrence of thick saturated regolith. However, there is no correlation between total borehole depth and yield ($r = 0.20$); the higher yields cannot be attributed to greater hard rock penetration. Thus, it can be established that to drill more deeply into hard rock where regolith is thick, as has evidently been the practice, is not cost-effective.

There was some doubt at the outset about the worth of subdividing the African surface into minor erosion surfaces. Although relative relief emerges as the strongest terrain criteria for borehole site selection, nevertheless the contrasts in mean yield on the different minor surfaces are significant. The reasons for this are not understood, as no systematic study was undertaken of mineral or chemical variations in profiles from the various surfaces. Some semblance of a pattern of variation is evident in the case of holes in the lowest relative relief category, with lower yields associated with the middle altitude surfaces, but no patterns are evident for higher relief categories. With or without patterns and without understanding of the reasons for the variations, the subdivision of the African surface appears justifiable in that it provides a second means of prognostication as concerns yields in particular situations. Thus, equipped with only a 1:50 000 topographic map from which altitude and relative relief of a particular locality can be easily established, and given the data on mean yields associated with different relief classes on different surfaces, shown in Table 5 and Fig. 4, a hydrogeologist or driller is able to assess the local site options. For example, in the case of a village at a particular altitude, the first criterion would be land with the lowest relative relief and the choice then becomes 'flat land on the surface, above it or below it?', a choice which can be made on the basis of the tabulated mean yields for the erosion surfaces. It must be emphasized, however, that the erosion surface criteria can only be applied to this study area. Elsewhere, deformation or tilting of the surface or surfaces may occur, so that absolute altitude can no longer be applied as a means of locating position in relation to specific surfaces.

The emergence of a distinct pattern of yield variation associated with relative relief categories and of different yields on different erosion surfaces indicates that here the influence of lithology is weak in comparison with that of geomorphology and leaching history. This is more than encouraging to the search for criteria for site selection in view of the inaccessibility of lithological information. However, again it must be stressed that this generalization is only applicable to this study area. In drier regions, for example, or on the younger erosion surfaces where there has been less time for geomorphological overprinting on the influence of lithology, rock type often emerges as a dominant influence on the extent to which weathering has penetrated and there are often significantly different yields associated with different lithologies (Wright 1988).

This study assists with regional assessments of hydrogeological potential, purely from relative relief and altitude. The low yields of the Dowa West and Lilongwe NE areas are consistent with their geomorphological contexts. Had the recognition of the particular importance of selecting sites in the lowest relief category been available at the time when these projects were undertaken, this could possibly have assisted. Following this analysis, future drilling programmes are rather better placed in terms of general hydrogeological expectations and in areas where low relative relief sites cannot be found proximate to the centre which requires a supply, then it may be appropriate to opt for holes which penetrate the hard rock.

It has already been established with some security that the shedding of precipitation from inselbergs facilitates deep weathering around their base (Ruxton 1958; Thomas 1974; McFarlane 1989c). In the case of this study area, however, there are indications that even higher yields are achieved on very flat land. Thus the better choice of a borehole site, to serve a centre near an inselberg, would be on flat land away from the berg rather than close to it. Again this generalization only applies to this type of terrain. In drier areas or on Post-African erosion surfaces where weathering is generally thinner, sites closer to the berg may be preferable.

This analysis has revealed the practice of drilling more deeply into hard rock in situations where relative relief is low and regolith is thick. This practice is not cost-effective in terms of increased yield.

The time constraints placed on this study allowed only preliminary exploration of relationships. The quality of the data would clearly justify further analysis. For example, median rather than mean yield values may better characterize the variations with terrain. Inter-quartile ranges could allow an assessment of the variability within terrain units, providing at least a risk assessment and possibly some understanding of criteria for the avoidance of bad risk situations. There are already indications, for example, that between erosion surfaces the relief on the basal surface of weathering may be relatively high (Wright et al. 1988) and since thick saturated regolith is important, it could well be that in such situations geophysical surveys are appropriate to avoid the 'dry highs'. Addressing such questions is now a relatively simple matter and these are important issues if available funds and exploration assets are to be used most effectively.

There seems little doubt that the pursuit of such studies of existing borehole data is constructive and ongoing computerization of data, as drilling proceeds, preferably within a systematic framework more specifically designed to maximize the information value, would allow valuable consolidation and refinement of site selection criteria.

Conclusions

The results of this statistical study provide a means of assessing regional hydrogeological potential from terrain type on the generally subdued African erosion surface in central Malawi. Very low relief areas, with few streams and extensive areas of dambo indicate thick total and saturated regolith and relatively high yields. Conversely, in areas of higher relative relief lower yields can be anticipated. This allows better scope for decison-making as concerns the type of borehole appropriate to the terrain and the needs of those to be served. For example, for a large group of consumers in a relatively high relief area the appropriate choice may be a hole which penetrates the saprock and hard rock rather than one which bottoms on the basal surface of weathering. Although the difference in the mean yields of these two types of holes is not great, and this appears to indicate the sustained openness of fractures and the maintained efficiency of bypasses like quartz stringers in the saprolite, nevertheless mean yields are some 20% higher where the saprock and hard rock are penetrated. This may be sufficient to make the difference between success and failure for a particular water need.

This study also makes some contribution to the question of whether or not a geophysical survey is cost-effective in terms of increased success rate. Although no systematic examination was made of holes drilled with and without geophysics, the emergence of clearly systematic variations in yield with terrain indicates that the influence of lithology is secondary to leaching history. At this point in any case it is not yet known which rock types are most favourably targeted. A geophysical survey which may identify places where rocks are more susceptible to weathering, and hence the regolith is thicker, in terrain which has generally thick regolith seems of doubtful value. Nevertheless, where relative relief is higher and an adequate thickness of regolith is in doubt, geophysics may allow a particular site to be identified as undesirable. Thus it may be that geophysics can be most effectively employed to eliminate from further consideration situations where there is the risk of thin regolith.

Given the constraint that a borehole must be close to a demand point, this statistical analysis helps to locate sites in generally low relief terrain where topographic features are very subdued. It identifies the paramount importance of the lowest available relative relief. Further, it indicates that, given low relief site options at various altitudes, those below the local erosion surface are likely to be higher yielding. More specifically, with the tabulated mean yields on and between specific erosion surfaces, it is possible for a hydrogeologist or driller to select the best sites, provided only with a 1:50 000 topographic map showing altitude and relative relief in terms of numbers of contours crossing kilometre squares adjacent to the centre to be served.

The study also provides some comment on the question of whether or not inselberg-base sites are particularly favourable. It suggests that, in this type of terrain, very flat land at least one kilometre distant from the berg is generally likely to be higher yielding than sites close to the berg.

It should be emphasized that these conclusions are to be regarded as specific to the area of terrain analysed. Elsewhere on the African surface in Malawi, although relative relief may equally well express relationships between regolith thickness, saturated thickness and yield, deformation of the erosion surfaces is likely to render absolute altitude an ineffective indication of which erosion surface occurs to a particular locality and hence altitude cannot be used to key into the mean yield values associated with the specific erosion surfaces in this area. Similarly, the generalizations which pertain to this type of terrain are unlikely to apply to the African surface in other countries where it lies in different bioclimatic zones or on post-African surfaces where less time has been available for the effects of weathering to overprint those of lithology. Nevertheless, the study does show that, with this type of approach, existing borehole data may be usefully analysed to establish terrain/yield relationships which can be of practical value. Data processing was, in this case, by no means exhaustive and there is clearly a case for both further analysis and ongoing assimilation of borehole data in order to refine and extend the drilling guidelines herein suggested by this study.

This paper is published by permission of the Director of the British Geological Survey (NERC). The work was carried out as part of the BGS Basement Aquifer Project, funded by the UK Overseas Development Administration. The second author was also seconded to the Water Department in Malawi under UK Technical Cooperation Programme. The authors would like to thank Francis Msonthi, Robinson Kafundu and Linda Mauluka of the Groundwater Section of the Water Department in Malawi, for provision and checking of borehole information, and John Talbot of BGS for assistance with the statistical handling of the data. Digitization of terrain characteristics was ably assisted by Claire Haynes, Frieda McFarlane and Sara and Ann Dixon. The authors thank Dr D. J. Bowden for his helpful comments on this paper.

References

ALLEN, D. J. 1987. *Hydrogeological Investigations using Piezometers at Chimimbe and Chikhobwe Dambo Sites, Malawi, in August–September 1986*. B.G.S. Basement Aquifer Project Unpublished Report No. EGARP/WL/87/4.

BENSON, M. A. 1959. Channel slope factor in flood frequency analysis. *J. Hyd. Div., A.S.C.E.*, 85, HY4.

BULLOCK, A. 1988. *Dambo Process and Discharge in Central Zimbabwe*. PhD thesis, Southampton, UK.

—— in prep. Variation in river flow regimes attributable to dambos in Zimbabwe.

—— & McFARLANE, M. J. in prep. A model of the hydrology and hydrogeology of dambo regions in Zimbabwe and Malawi. *Dambos and Duricrusts Symposium, Second International Geomorphology Conference, Friedrichsdorf, FRG, September 1989.*

DRAYTON, R. S., KIDD, C. H. R., MANDEVILLE, A. N. & MILLER, J. B. 1980. *A Regional Analysis of River Floods and Low Flows in Malawi*. Institute of Hydrology Report No. 72.

EGGLER, D. H., LARSON, E. E. & BRADLEY, W. C. 1969. Granites, grusses and the Sherman erosion surface, southern Laramie Range, Colorado–Wyoming. *American Journal of Science* 367, 510–522.

KING, L. C. 1963. *South African Scenery* (2nd edn) Oliver & Boyd, Edinburgh.

—— 1967. *Morphology of the Earth* (3rd edn) Oliver & Boyd, Edinburgh.

LISTER, L. A. 1967. Erosion surfaces in Malawi. *Records of the Geological Survey of Malawi* 7 (for 1965), 15–28.

McFARLANE, M. J. 1986. *Interpretation of Weathering Profiles (Hydrogeology of Regolith)*. B.G.S. Basement Aquifer Project, Unpublished Report (Open File).

—— 1989a. Dambos—their characteristics and geomorphological evolution in parts of Malawi and Zimbabwe with particular reference to their role in the hydrogeological regime of surviving areas of African Surface. *In: Groundwater Exploration and Development in Crystalline Basement Aquifers*, Commonwealth Science Publication, **1**, 254–310.

—— 1989b A review of the development of tropical weathering profiles with particular reference to leaching history and with examples from Malawi and Zimbabwe. *In: Groundwater Exploration and Development in Crystalline Basement Aquifers*, Commonwealth Science Publication, **2**, 93–145.

—— 1989c. Erosion surfaces on ancient cratons—their recognition and relevance to hydrogeology. *In: Groundwater Exploration and Development in Crystalline Basement Aquifers*, Commonwealth Science Publication, **1**, 199–253.

—— & WHITLOW, R. in press. Key factors affecting the initiation and progress of gullying in dambos in parts of Zimbabwe and Malawi.

—— & LEWIS, M. in prep. Dambo distribution and configuration in parts of Malawi. *Dambos and Duricrusts Symposium, Second International Geomorphology Conference, Friedrichsdorf, FRG, September 1989.*

MEIGH, J. R. 1987. *Low flow analyses of selected catchments in Malawi and Zimbabwe.* Preliminary report, B.G.S. Basement Aquifer Project.

PEDRO, G. 1968. Distribution des principaux types d'alteration chimique a la surface du globe. *Revue de Geographie Physique et Geologie Dynamique* **10**, 457–470.

PURVES, W. D. 1976. *A Detailed Investigation into the Genesis of Granite-Derived Soils'* PhD thesis, University of Rhodesia, Salisbury.

RUXTON, B. P. 1958. Weathering and subsurface erosion in granite at the Piedmont Angle, Balos, Sudan. *Geological Magazine* **95**, 353–377.

SMITH-CARINGTON, A. K. & CHILTON, P. J. 1983, *Groundwater resources of Malawi. Govt. of Malawi Report, Lilongwe.*

STRAHLER, A. N. 1950. Equilibrium theory of slopes approached by frequency distribution analysis. *American Journal of Science* **248**, 800–814.

THOMAS, M. F. 1974. *A Study of Weathering and Landform Development in Warm Climates*, Macmillan, London.

THOMPSON J. G. 1969. *The Classification and Cultivation of Wet Land and Vleis*, Unpublished Paper, Ministry of Agriculture, Harare, Zimbabwe.

TURNER, B. 1985. The classification and distribution of fadamas in central northern Nigeria. *Zeitschrift fur Geomorphologie N.F.*, Suppl.-Bd. **52**, 87–113.

WRIGHT, E. P. 1988. *Basement Aquifer Project. Progress Report for 1987–88, B.G.S. Overseas Hydrogeology Report, April 1988.*

——, HERBERT, R., MURRAY, K. H., BALL, D., CARRUTHERS, R. M., MCFARLANE, M. J. & KITCHING, R. 1988. *The Final Report of the Collector Well Project, 1983–1988, B.G.S.* Hydrogeology Research Group Technical Report WD88/31.

From WRIGHT, E. P. & BURGESS, W. G. (eds), 1992, *Hydrogeology of Crystalline Basement Aquifers in Africa*
Geological Society Special Publication No 66, pp 155–182.

The hydrogeology of crystalline aquifers in northern Nigeria and geophysical techniques used in their exploration

J. R. T. Hazell,[1] C. R. Cratchley[1,2] & C. R. C. Jones[3]

[1] *Water Surveys (UK) Ltd., Suite 2, Bayer Building, Lower Bristol Road, Bath BA2 3DQ, UK*
[2] *University of Wales, School of Ocean Sciences and Centre for Arid Zone Studies, Bangor, Gwynedd LL57 2UW, UK*
[3] *Mott MacDonald International, Demeter House, Station Road, Cambridge CB1 2RS, UK*

Abstract. The large number of rural water supply boreholes sited and drilled in northern Nigeria during the ground water decade of the eighties has provided much data on the hydrogeology of the Basement regolith and has led to the development of new techniques in locating aquifers. Regolith aquifer characteristics and their dependence on geology and environment are reviewed with particular emphasis on lithology, weathering pattern and fracturing in Bauchi, Kano and Sokoto States. Specific capacity is shown to be related to both lithology and grade of weathering; dry season water levels and frequency of occurrence of successful holes are related to lithology. Water quality data for Kano State show some high nitrate concentrations.

The process of borehole siting begins with desk study (records and photo-interpretation), then field reconnaissance of lithology, structure, water points, topography and soils. EM traverses and resistivity soundings follow in targeted areas. Maximum regolith thicknesses are estimated from the vertical to horizontal coil (EM) response using computed model graphs for 40 m and 20 m coil spacings. VES measurements at these sites give regolith resistivities, hence some estimate of its suitability as an aquifer. In addition EM traverses have located steeply-dipping fracture zones, dykes and pegmatites and interpretation of these in terms of strike and dip has been aided by results from a computer-controlled modelling system.

Until quite recently, the Nigerian crystalline basement was regarded as a poor and unimportant source of water. Dug wells serving the small communities which make up most of the rural population, though largely adequate, could penetrate only the upper parts of the regolith. The first downhole-hammer water drilling rig to operate in Nigeria, in 1964, made it possible to drill rapidly into hard rock, and to intersect fractures; at the same time, Dempster in Benin (then Dahomey) was investigating 'master joints' in granite terrain. (Dempster, A. UNDP pers. comm. 1964).

Although the resistivity method was used by the Geological Survey of Nigeria to locate deeply weathered zones from the mid-1940s onwards, the restricted availability of equipment and operators meant that most dug wells in rural areas were located by common sense, i.e. on low ground and as far as possible from outcrops. The routine use of resistivity in the 1960s for optimum siting of slim downhole-hammer boreholes was augmented in 1976 when EM and VLF were used in Bauchi State to locate high-yielding boreholes in a 'master joint'. Until 1980, the number of boreholes scientifically sited was numbered in hundreds.

With the coming of the 1980s, thousands of slim boreholes with handpumps were constructed for World Bank funded agricultural development programmes to improve the quality of life for rural communities. These provided data which, together with the advent of high resolution equipment and computer based interpretation methods, have greatly enhanced our understanding of crystalline aquifers and their importance.

In northern Nigeria, crystalline basement occurs over an extensive tract which covers parts of Sokoto, Katsina, Kaduna, Kano and Bauchi States (Fig. 1). Abstractable water is found both in the variably weathered zone or regolith and in fractures, weathered joints and dykes (Fig. 2). In featureless terrain with little outcrop they can be delineated and evaluated by geophysical techniques. The well established vertical electrical sounding (VES) technique has been supplemented and, in some surveys, almost replaced by modern EM sounding and traversing techniques especially during the past decade (McNeill 1980). The technique is used to locate zones of deeper than average regolith. These may be fairly narrow (less than 30 m wide) following bands of easily weathered rock; linear, following weathered vertical joints; or more extensive areas of deeper weathering. In addition, and particularly where the regolith is shallow, useful steep dipping fracture zones are detectable with EM traversing techniques.

In addition to standard computer interpretations of VES and EM34 sounding data, rapid regolith depth estimates have been made from ratios of the different components of EM34 profiles, according to graphs developed for simple regolith models. These are particularly useful in the field for the location of subsequent VES positions. Interpretation of the negative anomalies indicative of dipping conductors has been aided by the use of the table top computer-controlled EM modelling system developed by D. T. Hopkins & D. A. Vaughan of Geodak Ltd.

While much reliance is placed on geophysical data, due weight is given to airphoto interpretation and field observations of topography, soils and outcrops. The lithology of the parent rock and its structural history largely determine the hydrogeological characteristics of the regolith.

This paper reviews the results of recent exploration and development programmes in Bauchi, Kano and Sokoto States. The relationships between hydrogeological conditions in the Basement Complex, weathering grades (developed for Kano State) and geophysical interpretation are discussed in the light of these results and of a recent programme in Kaduna State.

Geological and geomorphological background

Geology

The crystalline outcrop of northern Nigeria is bounded to the southeast by the Benue–Gongola Cretaceous basin, to the north and east by the Quaternary Chad Basin, to the west by the Cretaceous to Tertiary Sokoto Iullmeden Basin, and to the southwest by the Cretaceous Niger trough (Fig. 1a).

Table 1 shows the principal stratigraphic and tectonic events of northern Nigeria. In bold type are those events which relate to the development of the regolith. Most of the tectonic events are manifest in the sedimentary rocks but also affect the crystalline areas. Even in the most stable part of the crystalline outcrop around

Dambatta in Kano State, the existence of Cretaceous sandstone in a fold or graben points to activity of Cretaceous or later date. Warping of the (?) Miocene laterite surface, which can be traced across large parts of the northern part of the Basement outcrop, also affected the development of the regolith.

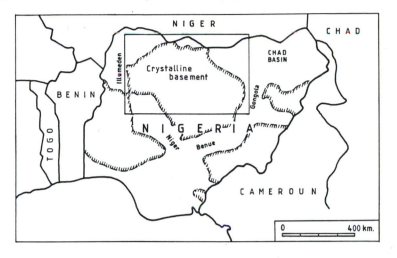

Fig. 1a. Location map showing extent of Crystalline Basement in northern Nigeria.

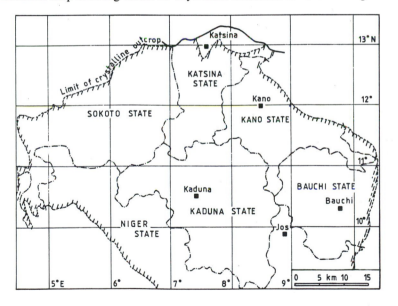

Fig. 1b. Part of northern Nigeria showing State boundaries within the Crystalline Basement outcrop.

Outcrops in the crystalline area give little indication of the relative frequency of the constituent rock types. Older and Younger Granites, quartzites and some migmatites are common in outcrop; schists, gneiss and metasediments generally are rare. In Kano State, an important aquifer, the pink felspar granite, is never seen in fresh outcrop. Borehole samples give a better indication of the frequency of occurrence of

Table 1. *Summary of geological history*

Age	Formation	Event	'African' axis: (N.5 to 20°E)	
			Movements etc.	Volcanics
Recent	Modern alluvium	Modern pluvial period		
Pleistocene	Dunes and fluvio-aeolian	Desertification	Alignment of middle Jama 'are/ middle Gongola rivers	
	Chad Formation	Chad basin fully developed.		
	Old alluvium	Erosion; drainage development. Alluvial infill in crystalline depressions/drainage channels.		
? ? ? Post-Miocene		*Master joints develop*	Joint axis	Jos and Biu Plateaux
		Warping	Warping axis	Cameroun chain
		Jointing of lateritic surface	Jointing axis	
Miocene	Continental Terminal,	Peneplanation, lateritization.		
Eocene	Kerri Kerri Formation, Gwandu Formation	Continued development of Chad and Sokoto basins; part of regolith covered and preserved		
Palaeocene		Lateritization; *some folding and shearing* parl. to Gongola arm	Shearing	
		Sokoto and Benue/Gongola basins		
Maestrichtian	Various esturarine and deltaic formations	Folding in the Niger and Benue/ Gongola basins	Folding of the Gongola arm	
Mid/lower Cretaceous	Marine and continental formations	Development of the Chad, Benue, Niger, Sokoto basins		Benue trough
		Massive erosion of crystallines		
Jurassic	Younger Granites	Emplacement of ring complexes	Complexes aligned	
	Older Granites	Emplacement of groups of complexes	Elongation of groups	
Palaeozoic-Pan-African		Intermediate and basic dykes	Axis of dykes	
		Long leucocratic dykes	Foliation	
Katangan	Metasedimentary formns.			

aquiferous rock types. In Kano State over three-quarters of boreholes are in granite or granitic rocks (though this is not always the country rock) (Table 2); in Bauchi State only 33% of boreholes were screened in granite.

Table 2. *Aquifer types and specific capacities, Kano State Aquifer*

| | Undifferentiated granite | | Metamorphic rocks | | Pink granite |
	Weathered	Fractured	Weathered	Fractured	
Number of boreholes	262	86	96	46	122
Percentage of total	43	14	16	7	20
Percentage successful	78	98	61	98	88
Of successful holes, percentage with air-lift yield over 100 litres per minute	15	18	12	22	17
Mean specific capacity (m^3/day/drawdown)	14.9	16.9	8.4	19.4	12

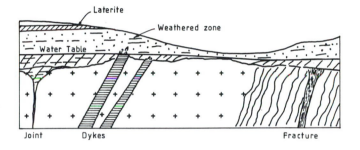

Fig. 2. Schematic outline of possible targets for water boreholes.

Development of the modern landscape

The evolution of the modern surface is closely related to the development of the crystalline regolith. Two aspects are relevant; the pattern of Tertiary and later warping with associated developments of joints and fractures, and the effects of Quaternary and later sedimentation on weathering and erosion.

Post-Miocene tectonic events. The laterite crust on the Kerri Kerri Formation (Continental Terminal) outcrop, tentatively ascribed to the Miocene, is a useful marker horizon which can be traced for considerable distances on to the crystalline outcrop of the Bauchi and Kano Plains. Undulations and joints in the surface, resulting from post-Miocene warping, are aligned N10E, parallel to the ubiquitous 'African' axis. The modern drainage in the crystalline areas can in places be demonstrated to have evolved from the undulations in the surface. Variations in the amplitude of warping may be used to indicate the probable intensity of joints in the basement, and hence the subsequent vulnerability of the regolith to further weathering.

Lowlands which are substantially lower than adjacent relics of the old surface reflect the amount of regolith which has been removed since the Miocene.

Quaternary and Recent accumulations. There is evidence that the essential pattern of the northeastwards directed drainage has changed little since before the Quaternary (Hazell *et al.* 1988), and fossil alluvium over 60 m thick has been mapped, often with a cover of fluvio-aeolian fadama silts. Patches of thick silt occur on uplands, notably on the Bauchi Plain. These sediments constitute superficial reservoirs capable of assisting recharge of the regolith and protecting it from erosion.

Groundwater environment

Disposition and development of the regolith

Accumulated data from many old and unsuccessful dug wells, and from geophysical traverses carried out during recent borehole siting indicate that in northern Nigeria the depth to fresh rock is on average only about 8 metres. This background is interrupted by areas of abnormally deep weathering, which are the principal targets for groundwater abstraction. The regolith has been divided into classified grades of weathering, from silty clay in which no relict textures are detectable (Grade IV), down to almost fresh rock (Grade IB, Fig. 3); and this sequence may be capped by a ferruginous crust with an associated zone of leaching. In parts of Kano State, where the laterite capping may be ascribed to the Miocene surface, it is reasonable to assume that much of the regolith developed prior to the Miocene, and indeed in these areas weathering is extensive, deep and highly developed. Elsewhere, as for instance in the Jama'are and Gongola valleys, much of the Tertiary regolith has been stripped by erosion, and the Grade II rocks are frequently exposed.

Table 3. *Mean depths of casing, drilling, pump setting, water level (metres)*

State	Group	Depth cased	Depth drilled	Pump setting	Rest water level
Bauchi	All holes		32.3	23.3	
Sokoto	All holes		36.1		
Kano	Undifferentiated				
	granite: weathered	36.7	42.8		12.0
	fractured	36.2	39.7		12.4
	Metamorphic:				
	weathered	43.9	54.8		16.1
	fractured	48.7	53.0		18.1
	Pink granite	35.6	40.7		15.2
Three States	All holes				12.2

The classification of grades of weathering which is of most practical use in assessing levels of the regolith as aquifers was developed for Kano State, based on descriptions of drill cuttings. An idealized section through a typical mature weathering profile is shown in Fig. 3. The weathering grades are those in common usage apart from IIA, which was introduced because a subdivision of stained rock was required. Groundwater is commonly found within the grades IV, III and IIa. Grades IV and III are not normally exploited because they contain clay and so have a lower permeability and produce dirty water relative to IIA. However, they are important as reservoirs, providing flow into grade IIA material. This is particularly relevant in areas where boreholes have been sited on narrow features, as illustrated in Fig. 4. Where water is found in grade II or IB material it usually follows discrete fractures.

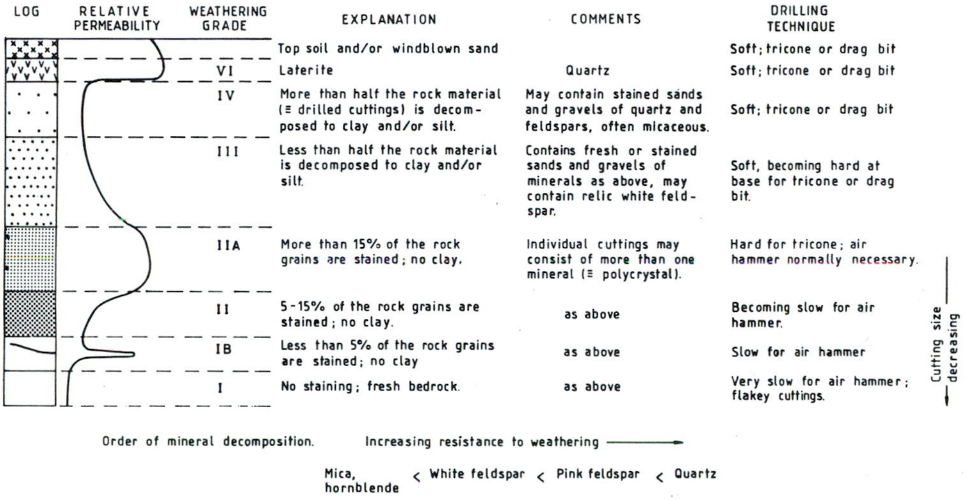

LOG	RELATIVE PERMEABILITY	WEATHERING GRADE	EXPLANATION	COMMENTS	DRILLING TECHNIQUE
			Top soil and/or windblown sand		Soft; tricone or drag bit
		VI	Laterite	Quartz	Soft; tricone or drag bit
		IV	More than half the rock material (≡ drilled cuttings) is decomposed to clay and/or silt.	May contain stained sands and gravels of quartz and feldspars, often micaceous.	Soft; tricone or drag bit
		III	Less than half the rock material is decomposed to clay and/or silt.	Contains fresh or stained sands and gravels of minerals as above, may contain relic white feldspar.	Soft, becoming hard at base for tricone or drag bit.
		IIA	More than 15% of the rock grains are stained; no clay.	Individual cuttings may consist of more than one mineral (≡ polycrystal).	Hard for tricone; air hammer normally necessary.
		II	5-15% of the rock grains are stained; no clay.	as above	Becoming slow for air hammer.
		IB	Less than 5% of the rock grains are stained; no clay	as above	Slow for air hammer
		I	No staining; fresh bedrock.	as above	Very slow for air hammer; flakey cuttings.

Cutting size decreasing

Order of mineral decomposition.   Increasing resistance to weathering ─────→

Mica, hornblende   < White feldspar < Pink feldspar < Quartz

Fig. 3. Weathering profile and grade classification, Kano State (MMP 1986).

It is common ground that the progress of weathering is controlled by external factors (climate, vegetation) and lithology plus tectonic influences. The relative importance of the two geological factors and their interaction is now better understood, though not completely encompassed.

The influence of lithology on aquifer characteristics

The depth and intensity of weathering of the regolith is not necessarily a measure of its usefulness as an aquifer. Instances abound of deeply weathered areas where the regolith is relatively impermeable. Deeply weathered schists and clay-rich meta-sediments in the western part of Kaduna State yield little water, whereas adjacent quartzites, scarcely weathered, but strongly jointed, may be successfully drilled. A deeply weathered biotite-rich rock among migmatites, adjoining an Older Granite ridge, at Galambi near Bauchi, yielded plenty of water, but so laden with weathered biotite as to choke the screen. The control exercised by rock types upon the aquifer characteristics of their weathering products is demonstrated by data from Kano State which show a clear difference in specific capacity between the classes of rock identified in drill cuttings as discussed later.

Adjacent lithologies need not vary greatly in composition or texture to give strongly different aquifer properties. This is exemplified by the varying response of different types of granite to deformation and subsequent weathering. The extent of kaolinization of felspars in granites largely controls the yield of boreholes and the quality of the water produced. In Kano State, pink felspar granites are dislocated by fracturing with only limited kaolinization, and give good yields; whereas adjacent areas underlain by white felspar granites weather to a putty-like consistency and suspended kaolin often vitiates the water abstracted. The response to forces on the two types of granite is observed in the upper Yedseram valley, near Mubi on the Cameroun border. A scarp exposes coarse porphyritic orthoclase granite beneath an almost identical coarse porphyritic plagioclase granite. Major joints in the orthoclase granite fade out when they reach the contact with the plagioclase granite.

Some cross-cutting rocks: steeply-dipping quartzites, pegmatites, Pan-African dykes etc., are dealt with under fractures below, since they combine strong lithological influence with localized fracturing. Other narrow, E–W Cretaceous to Recent vertical basalt dykes are invariably barren; joints in the fresh rock have no gape, and the rock transforms on weathering to a chocolate coloured clay.

Distribution, characteristics and significance of fractures

Two types of fracture are distinguished; one follows a recognizable horizon, often lithologically controlled, where the rock mass is dislocated, possibly a consequence of stress release; this forms a useful aquifer, for example at the base of the regolith, Grade IB (Fig. 4). The other is identified as a steeply dipping conductor on geophysical grounds or as a 'Master joint' following the 'African' axis.

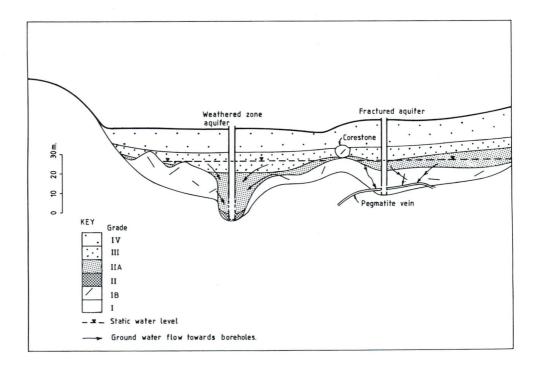

Fig. 4. Idealized cross-section through mature weathered zone (MMP 1986).

Steeply-dipping conductors may be the manifestation of banded biotite-rich gneisses (which are anathema to drillers but which can be recognized by the rapid alternation of geophysical response) or pegmatites or dykes. Pegmatites are commonly a locus of tectonic movement, and can be prolific conduits for groundwater; if they connect with a sizeable aquifer they can give sustained and outstanding yields. An intersection of two pegmatites near Bauchi yielded 300 litres per minute. A problem frequently met in Bauchi State is that major pegmatites often follow pan-African basic to intermediate dykes, which are themselves barren. The dykes are a wider target than the pegmatite, and a borehole can penetrate a long way into the barren dyke rock without meeting the narrow pegmatite. 'Master joints' are uncommon; they can extend for over 20 km as in the spectacular joint which runs through Bauchi town. This joint has no surface expression, is about 10 m wide, and was located on geophysical grounds alone.

In making a relative evaluation between larger areas of deep weathering, and steeply dipping fractures, the authors conclude that the former are generally preferred in northern Nigeria; though less prolific, they are easier to find and the geophysical data are less ambiguous. The main exception to this is in watershed areas with thin regolith.

Aquifer characteristics

The following data were analysed:

from Bauchi, Kano and Sokoto States, rest water levels, performance data (specific capacity and drilled depth);
from Bauchi and Kano States, lithological data;
from Kano State, more detailed data concerning grades of weathering;
from Kano State, water quality data.

The assessment of aquifer characteristics from well performance data is complicated by the interdependence of well design, the aim of each drilling testing programme, the siting techniques used, and the hydrogeological properties of the aquifers. This paper describes rural water supply boreholes in the three states where the same general criteria of siting, design and yield were adopted. All wells are constructed to a similar design with 100 mm ID UPVC casing with a 0.5 or 1.0 mm machine-slotted section for the screened portion. The diameter is suited to the hand pumps fitted to every well, which yield at best 15 l/m. The main variables in design are length of screen and total depth drilled. Many holes were overdrilled to allow for caving. In the overdrilled section, increases in yield were rare; it may therefore be concluded that differences in individual well performance are attributable to characteristics of the aquifer, and not to well design. Boreholes are sited in potentially the deepest and most permeable part of the regolith. Pump test data therefore do not reflect the characteristics of the entire aquifer, because of strong variations in depth and homogeneity. Borehole siting is confined to an area within a radius of 500 metres from each village centre; beyond this distance villagers will not normally travel to fetch water. For reasons of cost, the geophysical part of the investigation looks no deeper than 30 metres below static water level. It follows that better yields may be attainable beyond the limits thus imposed.

Rest water levels

Long-term monitoring of water levels was not undertaken as part of the agricultural projects. A few short-term hydrographs are available for part of Kano State for 1985/86, but in general rest water levels were measured only once, on completion of each borehole.

If during every month of the year, the locations of boreholes drilled were randomly distributed, the trend of the monthly average water level would indicate fluctuation due to seasonal influences. However, in the period July to October, when the land is wettest, drilling was not randomly located but took place on higher, drier ground, so that only during the period November to June can a significant trend be discerned. The distribution of depth to water with frequency has been plotted for the months of May (June is very similar) and January, when the natural depletion of the regolith is complete (Fig. 5). The total fluctuation is greatest around the mean water rest level (12.3 m) at about 5 metres and decreases at greater and lesser depths. In only 2% of the entire sample is the water rest level more than 24 metres; there is little variation in pattern from state to state.

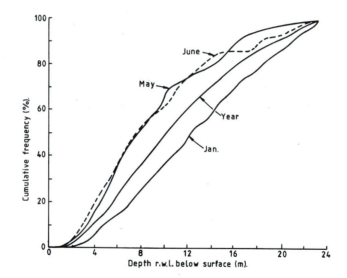

Fig. 5. Frequency of occurrence versus depth to water (at the time of completion) in 1318 crystalline boreholes in Bauchi, Kano and Sokoto States, Nigeria.

The overall mean rest water levels are included in Table 3. The data suggest that rainfall, which correlates roughly with latitude, is not the controlling factor on water levels; in Kano State most of the boreholes are within 0.75° of 12°N; in Sokoto State, most are within 1° of 10.5°N. Each sample is comparable in size; yet the mean for Bauchi State is 12.5% deeper than Sokoto and Kano combined. However, an inverse relationship is suggested between isohyets and rest water depth contours within Kano State, the deepest levels often being located in zones with less than 890 mm annual average rainfall. This apparent contradiction may be explained by the tendency in Bauchi State for villages to be located on high ground for defensive reasons; whereas in Kano State there is much less relief and a more tranquil history.

Table 4. *Water quality data for Kano State*

Parameter method	pH meter	EC meter	Hardness EDTA titration	Iron	Manganese colorimetric comparison	Nitrate	Fluoride HACH DREL/4 L/5
Range	—	0–10 000	—	0.50	0–3.0	0.4–50.0	0–2.0
Interval	0.01	2%	1.0	0.1	0.1	0.1	0.1
Limit adopted	5.0–9.2	2300	1125	3.0	1.0	11.3	1.5
Measured range	4.9–7.7	20–840	—	0–9	0–11	0–45	0–2.5
No. outside limit (500+ in sample)	7–5.0 0–9.2	0	0	21	36	30	3

However, in Kano State rest water levels in metamorphic terrain are deeper than average (Table 3), reflecting the fact that metasedimentary rocks usually underlie higher ground.

Previous workers (MMP 1986; Carter *et al.* 1963; Jones 1960) have commented on the interaction of rainfall, land use, evapotranspiration and water level, mainly in relation to sedimentary aquifers, but no reliable measurements of recharge or specific yield have been made.

Rest water levels are in equilibrium with their environment, and study of rest water levels is more relevant to well design and assessment of saturated thickness of the crystalline regolith than to estimation of resource potential.

Specific capacity

Specific capacity is more reliable than airlift yield as a measure of aquifer characteristics since it is not affected by uncertainties due to air–water ratio or submergence ratio. The three data sets are derived from tests with the following features:

State	Duration (hours)	Maximum yield* (l/min)
Bauchi	4	92
Kano	3	80
Sokoto	4	130

* imposed by the type of test pump available.

Comparison of specific capacity data for groups of wells requires a clearly defined treatment of boreholes abandoned owing to inadequate yield. These may be excluded altogether from the analysis thus further reducing the representativeness of the data for the whole aquifer. Alternatively, the abandoned wells can be ascribed a specific capacity lower than the lowest measured value. This approach is most useful when evaluating the siting technique.

In the entire sample of 1517 crystalline boreholes there is a big spread of specific capacities, and a mean value has little significance. There is a close correspondence in frequency distribution in the three states as illustrated in Fig. 6. 75% of all boreholes have specific capacities of over 1.5 litres per minute per metre of drawdown; ($2.2 \, m^2/$ day) so to achieve a village water supply yield of 15 litres per minute the drawdown would be less than 10 metres.

Lithology and fracturing

The relationship between lithology and specific capacity of the regolith in Bauchi and Kano States was analysed. In areas where sedimentary rocks overlie the crystalline regolith, the sediments inevitably dominate the overall aquifer characteristics. In tabulating data, boreholes thus affected have where possible been excluded from data sets.

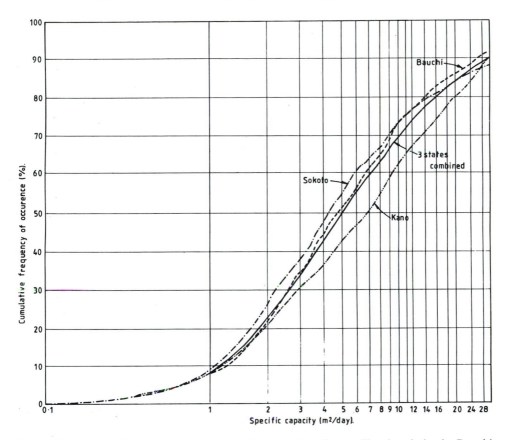

Fig. 6. Frequency of occurrence versus specific capacity of crystalline boreholes in Bauchi, Kano and Sokoto States, Nigeria.

Lithological data relate to the producing section of each borehole. The country rock, observed in the field or in non-aquiferous parts of the borehole, may be quite different.

In Bauchi State, three rock types are distinguished: 'granite', 'gneiss' and 'crystalline rock'. The last is thought to include migmatites, pegmatites, basic dykes and quartzitic rocks. Aquifers screened in each host are respectively 33%, 33% and 30%; the distribution of specific capacities in each is similar. This may be a consequence of the grouping of lithologies, since the better defined data from Kano State shows a clear difference.

In Kano State, aquifers were classified as shown in Table 2. Fig. 7 is a probability plot which relates specific capacity to lithology in Kano State. The sample takes into account abandoned boreholes, which were not tested.

Drilled depths

As mentioned above, there is no significant relationship between drilled depth and specific capacity, or drilled depth and yield. Drilling generally stops once the required depth for well construction and the required yield of 15 litres/min are achieved. While drilling in only slightly weathered rock, the 'make' of water increases in surges so that the margin of achieved yield over requirement again varies.

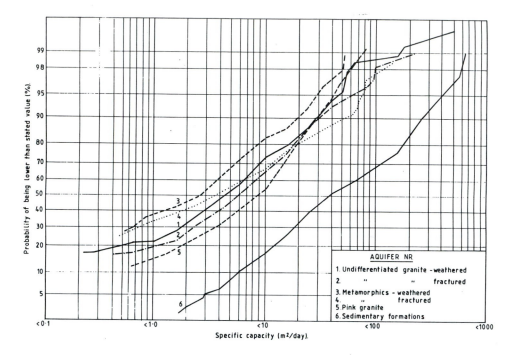

Fig. 7. Probability plot of specific capacity related to lithology Kano State, Nigeria (after MMP 1986).

Grades of weathering

An analysis of data from Kano State indicates that rock type influences depths of drilling. Aquifers in metamorphic rocks occur deeper below water level than weathered zone aquifers though owing to their high permeabilities total drilled depth is less.

Analysis of these data brings out a very strong relationship between specific capacity and grade of weathering screened (Fig. 8). Grade IIA is seen to be that most frequently screened (51%) and has the highest specific capacity (Fig. 9), between 1.5 and 2 times the average for all grades.

The depths to this horizon when first met are:

Kano State	Mean (m)	2/3 sample within (m)
Zone 1	18	9–26
Zone 2	26	16–31
Zone 3	17	9–23
All	20	9–26

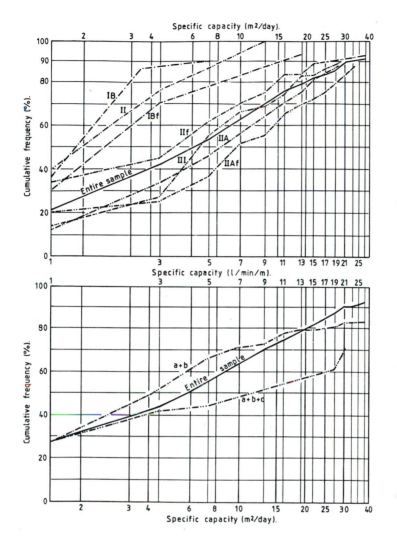

Fig. 8. Grade of weathering related to specific capacity of 460 crystalline boreholes in Kano State, Nigeria. Top: One aquifer only screened. Bottom: More than one grade screened. (Values over 27 l/min/m, not plotted, 8% of whole sample.) (a + b = 2 grades screened; a + b + c = 3 grades; f = fractured) (For description of grades, refer to text.)

In geographical Zone 1 of Kano State, the mean thickness of Grade IIA is 12.8 metres, where the next underlying layer (II or IB) has been identified. Where the borehole ends within or at the base of grade IIA, i.e. where the underlying layer is not identified, the mean is greater than 14.7 m. This is therefore a target layer of significant thickness and usefulness which is recognizable.

A comparison was made between the 'granitic' aquifers, where 54.6% of the sample was screened in Grade IIA, and 'metamorphic' aquifers where 40.6% are in grade IIA.

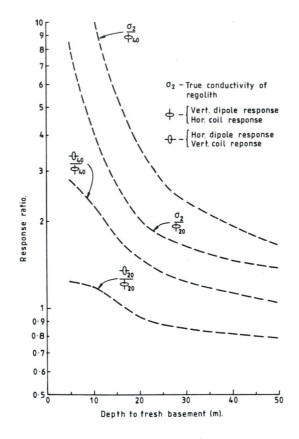

Fig. 9. EM response curves for regolith.

Water quality

All boreholes in Kano State were sampled at the end of the three-hour pumping test and the water quality assessed by measurement of the parameters shown in Table 4. The limits adopted, the method of measurement and the number of sites exceeding the limits are also shown.

Sites exceeding limits were selected for follow-up testing once handpumps were installed and recommendations for long-term monitoring made where necessary. It was found that iron and manganese levels tended to fall with borehole use. Nitrate concentrations in nearby dug wells were often found to exceed those in boreholes. 20 high nitrate sites in the southeast of the State were studied in detail but no single cause was identified to explain nitrate concentrations. It was concluded that some high nitrate values correlated with high EC values and that soil bacteria were a likely source of nitrates.

Many dug wells and boreholes exploit a similar part of the weathered zone and so the water chemistry is likely to be similar. In contrast, the microbiological quality of a properly constructed borehole is likely to be higher than in an open well. In principle, production zones deep below the water table or in sections of unusual lithology could produce water to which the consumers had not been traditionally exposed although little published work on trace metals and minor constituents exists. Indeed, few data on water quality are available other than in Du Preez & Barber (1965).

Table 5. *Resistivity and conductivity ranges for weathering grades*

Type of regolith/ rock	Provisional grade	Resistivity range (ohm m)	Conductivity range (mmhos/m)	Lithology
Silty-clayey regolith	III/IV	< 50	> 20	Mica rich biotite gneiss
Fine/medium- coarse regolith	III	50–100	20–10	Granite gneiss migmatites
Granular regolith	IIA	100–170	10–6	Older Granite coarse grained
Slightly weathered regolith/fractured rock	II	170–270	6–4	All
Fractured rock	II/I	> 270	< 4	All

Geophysical assessment of crystalline aquifers

Following desk study (researching of records, identification of potentially useful photolineaments) target areas for geophysical siting are located by standard field procedures (geological, hydrogeological, topographic and soils reconnaissance). To locate and evaluate regolith and fracture aquifers, a combination of VES and EM has proved to be the most effective and is becoming a standard method of exploration (Jones 1985; Jones & Beeson 1985; Beeson & Jones 1988; Palacky, Ritsema & de Jong 1981; Reynolds 1987; Hazell, Cratchley & Preston 1988). A schematic outline of the possible targets for water borehole sites in Basement areas is shown in Figs 2 & 4. The advantages of the combined EM/VES approach include the following:

(a) The two coil, Slingram type EM method is used as a rapid reconnaissance tool to locate areas of deeper weathering or steeply-dipping fracture zones. This requires no physical contact with the ground, is faster, and gives more diagnostic information than resistivity traversing at constant separation.

(b) From measurements with the two different coil orientations (vertical and horizontal dipoles) and at 20 m and 40 m coil separations, reasonably good estimates of depth of weathering and true regolith conductivity can be obtained from ratios of the two dipole values at each coil separation. These allow preliminary BH sites to be chosen on the basis of suitable depth and appropriate conductivity values (see below).

(c) VES measurements are limited to these chosen EM sites and aligned parallel to joint or foliation direction that is apparent from outcrop or indicated by EM trends across parallel lines. Rapid VES interpretation in the field using curve matching or approximate longitudinal conductance estimates from the plotted field curves (see below) can be compared with the EM interpretation and sites marked at the time of the field work.

(d) Computer interpretation of both VES and/or EM data (sounding) can be used to obtain a ground resistivity model which best fits the two sets of data. This can also be approximated from longitudinal conductance values. The joint computer interpretation helps to reduce ambiguity caused by equivalence (Hazell *et al.* 1988; Van Kuijk, Haak & Ritsema 1987).

(e) The EM traversing technique can also be used to locate accurately steeply dipping fault/fracture zones, pegmatites and dykes. In watershed areas of Bauchi State where the regolith is shallow, such fracture zones form the only potential source of groundwater. They require accurate delineation and to a limited extent this can be done with closely spaced EM stations and in suitable circumstances estimates of dip and strike can be made.

Geophysical criteria and strategy in deeply weathered regolith

The geophysical target is a reasonably thick and extensive area of saturated weathered rock. Taking into account mean values of specific capacity and drain-out depth to water, the minimum depth of weathering has to lie between 20 m and 30 m in order to provide the required yield throughout the dry season and an acceptable success rate. In addition, granular aquifers are preferred to clayey and silty ones, combined with a preference for slight weathering and fracture at the base of the regolith. This is the equivalent of locating Grade II or IIA beneath Grade III.

Geophysical criteria—general. Resistivity/conductivity criteria related to lithology and weathering have been established by Beeson & Jones (1988) for Kano State, by Hazell *et al.* (1988) for Bauchi State and by Hazell and Cratchley during current work in Kaduna State. These are reasonably consistent although the Beeson & Jones criteria allow for variations in groundwater conductivity, whereas the current work assumes these to be below 100 μmhos/cm. Resistivity soundings indicate three, four or five layer cases with one or more surface dry layers of high resistivity, including lateritic ironstone at up to 5000 ohm m, underlain by a relatively low resistivity weathered zone with a range of values of 30 ohm m (33 mhos/m) to 270 ohm m (4 mhos/m). This range generally reflects the preferred lithological range, high values within the range representing granular regolith, intermediate values silty clayey regolith or more saline groundwater. The zone of fracturing at the base of the regolith is very difficult to detect geophysically unless of greater thickness than the overlying regolith itself when the final leg of the resistivity curve ascends at less than the slope of 45 (the latter indicative of fresh massive rock of effectively infinite resistivity). In current work, five categories of resistivity/conductivity are assigned for comparison with lithological logs from drilling (Table 5). They reflect previous experience in Bauchi and Kano States and are reasonably consistent with values quoted in Table 1 of Reynolds (1987) for Kano State.

Geophysical strategy in deeply weathered regolith. For the ranges of resistivity given above, EM34 responses for simple three-layer models have been computed with the 'low induction number' equations of McNeill (1980) for regolith depth ranging between 10 and 50 m and for 20 and 40 m coil separation. A top dry layer of 4 m thickness is assumed in the model corresponding approximately to a water table depth of about 8 m which is common in northern Nigeria. Values of the response

Table 6. *Steeply dipping features, Bauchi State*

Geological feature	Strike	Possible dip	Width (max)	Groundwater potential
Pan African dyke (dolerite)	NW–SE WNW–ESE	45°–90°	10m	Good where disturbed and reintruded. Can be multiple
Tensional fracture zone (some qtz filled)	SSW–NNE	60°–90°	3m	Good. Fairly open structure
Tertiary dyke (basalt)	E–W	90°	1m	Very poor. Tight, little disturbance
Pegmatite	?	40°–90°	3m	Good. Variable in geometry

ratios (RR), i.e. horizontal dipole response/vertical dipole response (θ/φ, 20 m and 40 m cables) are diagnostic of regolith depth according to curves such as those in Fig. 9 which result directly from the computation. These are virtually independent of the regolith conductivity, σ_2 which can be calculated from the factor read off from the second set of curves and multiplied by the measured $\varphi40$ or $\varphi20$ value. Thus from the field EM data at 40 m and 20 m coil spacings, it is a simple matter to calculate the ratios, read off the estimated depths of regolith and calculate the true conductivities corresponding to each of the ratios. In general these values do not agree because of departures from the model, particularly multi-layers and the thickness of the upper dry layer, but mean values for both depth and conductivity appear to give acceptable first approximations ($+/-10$ metres) to the true values when compared with VES sounding results and limited drilling data. Most importantly, maximum depth of weathering along an EM traverse is indicated by high $\varphi40$ values and appropriate ratios, and potential borehole sites are marked on this basis, normally to be followed by a VES measurement at appropriate alignment (e.g. along an EM feature or parallel to known joint or foliation direction). However, the two examples in Fig. 10 show boreholes sited successfully on the basis of EM 34 alone. At Num Gagara, the estimated 24 m depth of regolith of 31 mmhos/m conductivity is compared with a logged depth of 28 m of highly to moderately weathered biotite gneiss overlying fresh biotite gneiss. At Badagari, the ratios indicate a weathered depth of greater than 50 m and conductivity of 30 mmhos/m. The BH log showed 0–3 m laterite, 3–54 m weathered biotite granite. Both holes were successful.

This method, particularly when combined with carefully sited and aligned VES measurements is rapid in both execution and field interpretation, particularly as Basement type VES curves with 45° final ascending leg are themselves amenable to rapid interpretation using the 45° line to estimate longitudinal conductance of the regolith (Kunetz 1966; Zohdy *et al.* 1974).

Geophysical criteria and strategy for steeply dipping fractures

In northern Nigeria, steeply dipping fracture zones are considered as potential borehole sites where the saturated weathered zone is < 10 m thick and this occurs for example in certain watershed areas of Bauchi State. These features which may be quite narrow (5 m) can be located using the EM34 from their characteristic negative anomaly (Fig. 11) on the vertical dipole response.

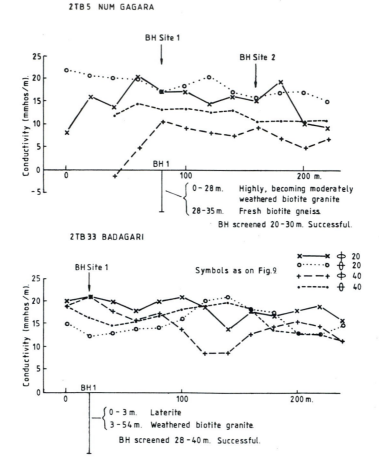

Fig. 10. Recorded EM responses and borehole results at Num Gagara and Badagari, Bauchi State.

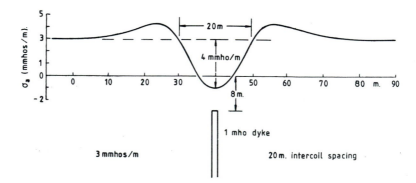

Fig. 11. Vertical dipole response to a vertical dyke (McNeill 1980).

The two coil EM (Slingram type) system can give diagnostic information about the dip and width of such structures (Fig. 12). Examples of theoretical and model experiments have been published (e.g. Ketola & Puranen 1967; Ketola 1968). A computer controlled desk top model EM system which can simulate a variety of coil orientations and geometries of the conductive fracture zone has recently been developed by Hopkins and Vaughan (see schematic Fig. 20). While it has not been possible to simulate accurately the true and relatively low conductivity values for a geological type conductor (as opposed to a mineralized type conductor), nevertheless, in a semi-quantitative way it has been possible to obtain useful diagnostic information on the geometry of dipping conductors (width of zone, dip and strike) from the anomaly shape on both horizontal and vertical dipole coil orientations.

TYPICAL RESULT FROM OUR MODELLING STUDIES

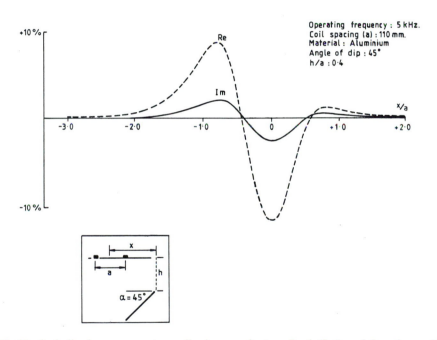

Fig. 12. Vertical dipole response to a dipping conductor. Real (Re) and Imaginary (Im) components.

EM criteria for geometry of dipping fractures. Previously published model responses (Ketola & Puranen 1967; Ketola 1968) and work by Hopkins for this paper with the computer controlled model EM system which he developed for narrow conductive sheets at different angles of dip and strike lead to the following diagnostic features:

(a) Horizontal coil (vertical dipole). A positive shoulder on the 'up dip' side of the conductor can be used to estimate an approximate dip angle provided the background levels are relatively undisturbed (Fig. 12). The negative anomaly should have a width equal to the coil spacing if the conductor is narrow. If wide or multiple, the anomaly width increases approximately by the width of the conductive zone.

(b) Vertical coil (horizontal dipole). This system is very sensitive to change in azimuth relative to conductor strike. At 90° to the conductor, the anomaly is a reduced amplitude version of the vertical dipole response. At fairly acute angles of intersection, the response becomes strongly positive and dip has less effect (Fig. 13). To a limited degree, these characteristics which apply to both in phase and quadrature responses can be used to obtain an indication of both dip and strike of a conductive fracture zone.

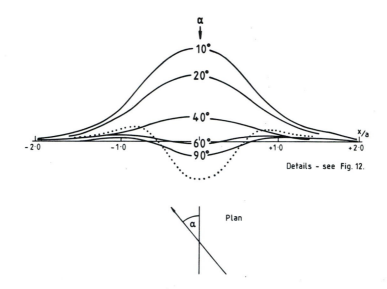

Fig. 13. Horizontal dipole response to a dipping conductor at varying azimuths, *a*.

EM strategy for steeply dipping fractures. Traverses with either EM34 (20 m and 40 m coil separation) or Maxmin (50 m or 100 m coil separation) can be used to locate and define negative anomalies. If the feature is to be accurately positioned and the anomaly shape is to be used as diagnostic of dip, strike or width of the zone, then a maximum station interval of one quarter of the coil separation, i.e. 5 m at 20 m coil separation or 10 m at 40 m coil separation should be employed over the anomaly itself which can first be located at the standard station interval used for deep regolith (20 m station interval throughout). By comparison with model experiment traverses with, for example, the computer controlled desktop EM system used in the current work, a qualitative or semi-quantitative assessment of the geometry of the conductor can be made.*

In Bauchi State, steeply dipping conductors may be produced by various geological features, not all of which form suitable targets for water boreholes. They can be identified in aerial photographs (see for example Fig. 14) and distinguished mainly on the basis of strike direction which is a very important diagnostic feature for

---

* In general, responses on the EM34 system are equivalent to out-of-phase responses in model experiments (also termed 'quadrature' or 'imaginary' (Figs 12, 16).

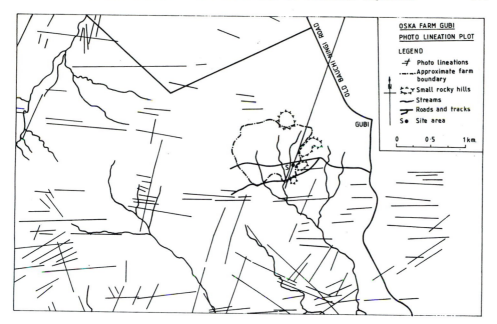

Fig. 14. Photolineation plot, Gubi Farm, Bauchi State.

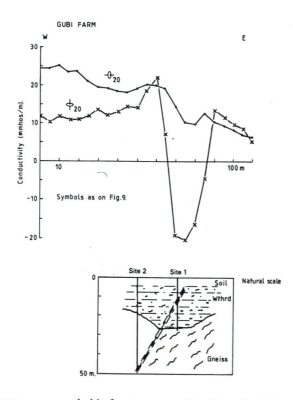

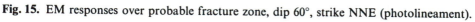

Fig. 15. EM responses over probable fracture zone, dip 60°, strike NNE (photolineament).

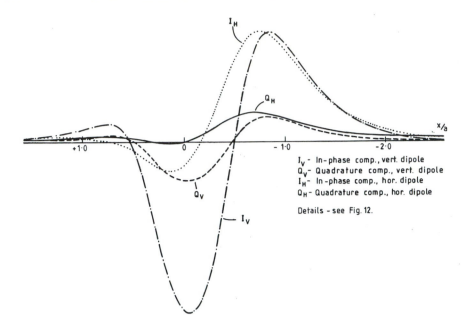

I_V - In-phase comp., vert. dipole
Q_V - Quadrature comp., vert. dipole
I_H - In-phase comp., hor. dipole
Q_H - Quadrature comp., hor. dipole

Details - see Fig. 12.

Fig. 16. EM model results, dip 45°, azimuth 90°.

groundwater potential. They are summarized in Table 6 and examples of the geophysical responses are discussed below.

(a) Gubi Farm, located on a photo lineament with NNE strike (Fig. 14). Both $\theta 20$ and $\varphi 20$ responses (Fig. 15) are consistent with a dip of 60° − 70° and strike almost perpendicular to traverse from comparison with model results (Fig. 16). Note also the change in regolith conductivity from west to east. The high $\theta 20$ values indicate fairly high conductivity at less than the 4 m depth in the model used for Fig. 9. This is a NNE-striking fracture zone on which borehole 1 was dry but borehole 2 struck water in the fracture zone.

(b) Siyi. All four responses are consistent with a 50° dip perpendicular to traverse indicating a strike of WNW–ESE. The $\theta 20$ data suggest a change in regolith conductivity across the dyke. The width of the conductive zone is about 10 m which may indicate a multiple dyke of the Pan African type. Borehole 1, drilled too close to the negative was dry but borehole 2 down dip was successful. The former encountered dolerite (as indicated in Fig. 17) but the second stopped short of the dyke and its success may be due to the more weathered basement on the up-dip side.

(c) Bungun. Very unusual EM responses were recorded at Bungun (Fig. 18) with a very high $\theta 20$ response and $\varphi 20$ response apparently a negative with pronounced positive shoulders and the whole anomaly shifted vertically to correspond with a background conductivity of 10–15 mmhos/m. Figure 19 indicates the closest qualitative model result to the field curve shapes, i.e. a 45° dip and the traverse line at 30° to the strike of the feature. The borehole encountered a weathered zone to 40 m depth and then weathered basalt from 40 to 50 m. The hole was successful and the most probable interpretation is a dipping and fairly wide Pan African dolerite dyke, at a low angle of strike to the traverse direction.

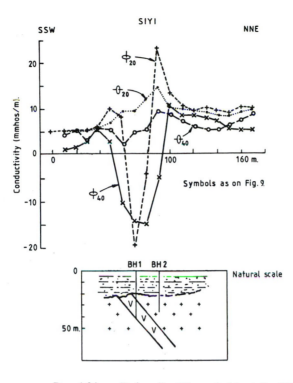

Fig. 17. EM responses over Pan African Dyke, dip 50°, probable strike WNW.

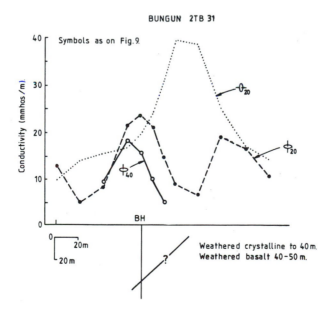

Fig. 18. EM responses over Pan African Dyke, dip 45°–50°, traverse intersects strike direction at 30°.

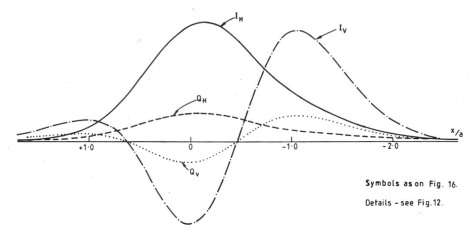

Fig. 19. EM model results, dip 45° (to right), azimuth 30°.

Conclusions

The crystalline basement outcrop of northern Nigeria has areas of deeply weathered regolith which consititute useful aquifers, separated by larger areas of relatively barren shallow regolith. Aquifers are variable in extent and thickness, ranging from broad areas of deep weathering to narrow joints and fracture zones. The material which constitutes individual aquifers is varied in composition, granularity and hydraulic characteristics, depending on the nature of the host rock, its tectonic history and the development of weathering. Lithology of the host rock influences the nature of the regolith, and both lithology and tectonic influences control the disposition of deeply weathered zones and of fractured zones. A weathering grade classification has been established for Kano State and individual grades correlate with aquifer yields. Borehole performance is dependent on the aquifer and not to any discernible extent on small variations in borehole diameter and depth. It is important to recognize that lithological data from boreholes, sited as they are in the most permeable and thickest regolith, are not necessarily representative of regolith lithologies as a whole.

Patterns in water level vary little geographically, the mean dry season water level is 12.2 m and depths to water rarely exceed 24 m. The seasonal fluctuation is 5 m or less. Distribution of specific capacities is uniform geographically and 75% of all boreholes have specific capacities of over 1.5 l/min/m of drawdown (approximately 2 m²/day). Within the weathered profile, fractured horizons towards the base of the regolith are the most productive and weathering grade IIA forms the most permeable horizon. A combination of EM34 and VES geophysical measurements has proved to be effective in locating and assessing regolith aquifers. Rapid estimation of depth of weathering from the vertical to horizontal coil response ratio and based on a simple computer model has proved reasonably consistent with borehole results and VES interpretations. In addition, EM traverses in watershed areas of Bauchi State have successfully located and established the approximate geometry of water bearing dykes, fracture zones and pegmatites from a comparison between recorded EM profiles and results from a computer-controlled analogue modelling system. Estimation of regolith lithology has been made from the interpreted resistivity values of

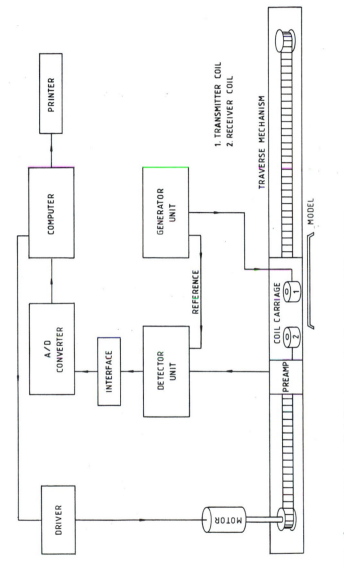

Fig. 20. Block diagram of computer-controlled EM dual coil table-top modelling system.

the weathered zone. To a degree, there is a correlation between ranges of values for resistive and weathering grades. However, the important Grade IIA horizon is unlikely to be located geophysically unless it is thicker than the overlying grades. Successful boreholes have been located on the basis of EM survey alone but in general, confirmation with VES measurements is preferred.

While geophysical data are dominant in aquifer prognosis and borehole siting, especially in featureless terrain with a few outcrops, the selection or rejection of a borehole location is also influenced by the type, frequency and structure of outcrops, the topography and nature of the soil and evidence from aerial photographs.

The authors are grateful to the Chief Irrigation Engineer, KNARDA, Kano, the Chief Engineer, BSADP, Bauchi and the Chief Engineer, SARDA, Sokoto for the use of borehole data. We also thank many colleagues and field workers who have contributed to this work, D. T. Hopkins for running the EM modelling experiments described in this paper and Derek Vaughan, Geodak Ltd, Reading for providing the modelling equipment.

References

BEESON, S. & JONES, C. R. C. 1988. The combined EMT/VES Geophysical method for siting boreholes. *Groundwater*, **26**, 1, 54–63.

CARTER, J. D., BARBER, W. & TAIT, E. A. 1963. The geology of parts of Adamawa, Bauchi and Bornu Provinces in north-eastern Nigeria. *Geological Survey of Nigeria Bulletin*, **30**, 108.

DU PREEZ, J. W. & BARBER, W. 1965. The distribution and chemical quality of groundwater in Northern Nigeria. *Geological Survey of Nigeria, Bulletin*, **36**, 67.

HAZELL, J. R. T, CRATCHLEY, C. R. & PRESTON, A. M. 1988. The location of aquifers in crystalline rocks and alluvium in Northern Nigeria using combined electromagnetic and resistivity techniques. *Quarterly Journal of Engineering Geology* **21**, 159–175.

JONES, C. R. C. & BEESON, S. 1985. The EM/VES Geophysics Technique for Borehole Siting, Kano State. *In:* OGUNTONA T. (ed.) *Advances in Groundwater Detection and Extraction*, International Conference on Arid zone hydrology and Water Resources, Maiduguri.

JONES, D. G. 1960. The rise in the water-table in parts of Daura RF7 and Katsina emirate, Katsina Province. *Rec. Geol. Surv. Nigeria 1957*, 24–26.

JONES, M. J. 1985. The weathered zone aquifers of the Basement Complex areas of Africa. *Quarterly Journal of Engineering Geology* **18**, 35–46.

KETOLA, M. 1968. *The Interpretation of Slingram (Horizonal Loop) Anomalies by Small-scale Model Measurements.* Geological Survey of Finland. Report of investigations No. 2.

KETOLA, M. & PURANEN, M. 1967. *Type Curves for the Interpretation of Slingram (Horizontal Loop) Anomalies over Tabular Bodies.* Geological Survey of Finland. Report of investigations No. 1.

KUNETZ, G. 1966. Principles of direct current resistivity prospecting. Geo-exploration Monographs Series 1—No. 1. Gebruder Borntraeger, Berlin 103.

MCNEILL, J. F. 1980. Electromagnetic terrain conductivity measurements at low induction numbers. Geonics Ltd. Technical Note TN6, 15.

PALACKY, G. J, RITSEMA, I. L. & DE JONG, S. J. 1981. Electromagnetic prospecting for groundwater in Precambrian terrains in the Republic of Upper Volta (Burkina Faso). *Geophysical Prospecting*, **29**, 932–955.

REYNOLDS, J. M. 1987. The role of surface geophysics in the assessment of regional groundwater potential in Northern Nigeria; *In:* CULSHAW, M. G., BELL, F. G. & O'HARA (eds). *Planning and Engineering Geology.* Geological Society, London, Engineering Geology Special Publication, 4, 185–190.

VAN KUIJK, J. M. J, HAAK, M. J. & RITSEMA, I. L. 1987. Reduction of equivalence in layer-model interpretation by combination of electrical resistivity soundings and electromagnetic conductivity measurements; some case histories in groundwater survey. *Journal of African Earth Science* **6**, 3, 379–384.

ZOHDY, A. A. R. *et al.* 1974. *Application of Surface Geophysics to Groundwater Investigations.* Geological Survey U.S. Geological Survey Chapter 01, 116.

From WRIGHT, E. P. & BURGESS, W. G. (eds), 1992, *Hydrogeology of Crystalline Basement Aquifers in Africa*
Geological Society Special Publication No 66, pp 183–201.

Borehole siting in an African accelerated drought relief project

R. D. Barker,[1] C. C. White[2] & J. F. T. Houston[3]

[1] *School of Earth Sciences, University of Birmingham, Edgbaston, Birmingham, B15 2TT, UK*
[2] *Aspinwall & Co., Walford Manor, Baschurch, Shrewsbury, SY4 2HH, UK*
[3] *Water Management Consultants, 1401 Seventeenth St, Suite 310, Denver, Colorado 80202, USA*

Abstract. In response to recent droughts in central Africa, an accelerated well drilling programme was commissioned in Victoria Province, Zimbabwe. Geophysical techniques involved in the siting of 370 boreholes included electromagnetic profiling and resistivity sounding, the latter using the offset sounding system. An analysis of the field data showed that although observed errors were low, a significant proportion of the soundings were affected by strong lateral changes in resistivity, this probably indicating the presence of a very irregular bedrock surface. Nevertheless, mean regolith thickness and lithology were reliably estimated.

Geomorphological features which were found to have the greatest potential for water supply boreholes include valleys and zones of weathering around bornhardts. In these situations geophysical surveys proved useful in locating areas of thickest regolith. Borehole yield could then be estimated from the measured formation resistivity although a precise relationship was somewhat masked by the clay mineralogy of the regolith and the type of bedrock which must be taken into account when siting boreholes.

Recent droughts in many parts of Africa have led to a large number of aid programmes aimed at providing water supply for rural communities. One such programme of drought relief was commissioned by the European Economic Community and the Government of Zimbabwe following failure of the rains in this country between 1981 and 1983. The work was carried out between September 1983 and May 1984 and involved the siting, drilling and testing of 370 boreholes and the installation of handpumps. The necessarily accelerated nature of the project meant that all aspects of the work had to be realized using fast, cost-effective techniques and in the borehole siting phase this meant that geophysical techniques were essential. This paper evaluates the use of the geophysics in this project paying particular attention to the role of resistivity sounding.

Victoria Province is located on the edge of the northern block of the Kalahari Craton and is an area of Precambrian igneous and metamorphic basement terrain (Fig. 1). Overlying most of the bedrock is a variable thickness of regolith formed by deep in situ chemical weathering over a long period of time. Thickness and lithology of the regolith depends upon bedrock type, amount of fracturing, subsequent erosion and climate. There is good evidence to suggest that greater age and rainfall lead to a thicker regolith. Fissure systems in the granite enhance weathering and may leave rounded interfissure blocks in the regolith known as corestones. In Victoria Province, the transition from weathered regolith to unweathered granite was observed to be relatively sharp, occupying only a few metres and often only a few centimetres. In gneiss, however, the transition is more gradual sometimes with the development of a zone of partial weathering many metres thick. In general the granite regolith is less thick than in gneissic areas.

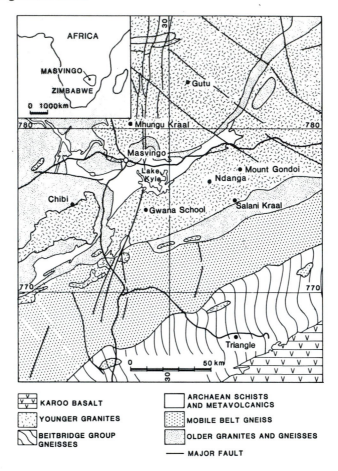

Fig. 1. Geology of the area around Masvingo and location of example surveys.

The regolith is largely composed of quartz grains, mica and clay minerals. Kaolinite is the principal clay mineral but in areas with less than 800 mm of rainfall, montmorillonite may be metastable. Leaching and eluviation of the clays help the regolith develop as an aquifer. The development of pipes greatly contributes to its permeability.

Igneous and metamorphic rocks such as those found in Victoria Province generally have very low primary porosity but the presence of faults and joints leads to the development of a fissure system which may give the rocks a high secondary porosity. The in situ weathering of these basement rocks may also serve to increase porosity. Thus there are two major modes of occurrence of groundwater in Precambrian Basement terrain: the weathered regolith and the fissured bedrock.

The regolith system is generally in hydraulic continuity with the bedrock system.

Consequently a borehole drilled into bedrock will often draw upon the storage in the overlying regolith. The regolith has a relatively high storage capacity but low permeability (Acworth 1987), whereas the fissured bedrock has a low storage capacity but a relatively high permeability (Fig. 2). Thus a borehole drilled into thick regolith might provide a low yield but more than enough to supply a continuously operated hand-pump, but one that was drilled through thick saturated regolith and into fractures in the underlying bedrock would provide a sustained, high yield.

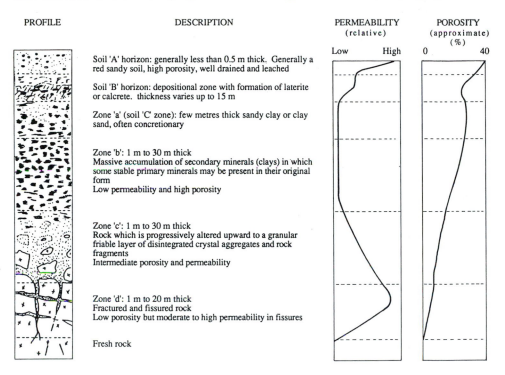

| PROFILE | DESCRIPTION | PERMEABILITY (relative) | POROSITY (approximate) (%) |

Soil 'A' horizon: generally less than 0.5 m thick. Generally a red sandy soil, high porosity, well drained and leached

Soil 'B' horizon: depositional zone with formation of laterite or calcrete. thickness varies up to 15 m

Zone 'a' (soil 'C' zone): few metres thick sandy clay or clay sand, often concretionary

Zone 'b': 1 m to 30 m thick
Massive accumulation of secondary minerals (clays) in which some stable primary minerals may be present in their original form
Low permeability and high porosity

Zone 'c': 1 m to 30 m thick
Rock which is progressively altered upward to a granular friable layer of disintegrated crystal aggregates and rock fragments
Intermediate porosity and permeability

Zone 'd': 1 m to 20 m thick
Fractured and fissured rock
Low porosity but moderate to high permeability in fissures

Fresh rock

Fig. 2. Vertical profile through regolith showing variation of weathering, storage capacity and permeability (after Acworth 1987).

The objective in using geophysical techniques is generally to locate a zone of thick regolith. Where the regolith is less than 10 m thick it may be possible to attempt to locate narrow fissure zones but generally in areas of thick regolith it is only possible to locate very broad, fractured and weathered zones below the regolith. However, it is reasonable to assume that the maximum regolith development occurs where the degree of fracturing within the bedrock is greatest and, therefore, boreholes sited at points of maximum regolith development have the greatest chance of locating fissures in the underlying bedrock.

General borehole siting philosophy in Victoria Province was based on these considerations although a lower limit of regolith thickness and lower and upper limits of resistivity (relating to clay content and permeability) of regolith were set and continuously reviewed through the duration of the project.

The accelerated nature of the project imposed severe limitations on the amount of geophysics which could be carried out by the siting team which was required to site an average of at least three boreholes per day. In this situation most of the geophysical surveys were used merely to confirm the suitability of a site selected on initial hydrogeological and logistical considerations. Due regard had always to be given to area of catchment, distance from village and availability of alternative supplies. For example, a good supply located much further from a village than a contaminated supply was sadly unlikely to be used. In this project around two thirds of the sites were positively confirmed as suitable with a small amount of geophysics while the remainder required a varying degree of additional geophysical survey before a suitable site could be located.

Clearly where the speed of measurement was of prime consideration only very field-efficient geophysical techniques could be employed. For this reason in this programme only resistivity sounding, electromagnetic profiling and, very occasionally, magnetic techniques were employed. The field techniques and results are discussed below.

Electromagnetic surveys

Electromagnetic measurements were carried out using a Geonics EM-34 ground conductivity meter, this instrument providing a direct reading of apparent conductivity in the region of the measuring coils. Three coil separations are available, each providing a different depth of investigation. If depth of investigation is defined as the median depth above and below which 50% of the signal originates (Edwards 1977), appropriate values for the three coil separations (10 m, 20 m and 40 m) are 3.8 m, 7.6 m and 15.2 m respectively for vertical coils and 8.7 m, 17.4 m and 34.8 m for horizontal coils. These figures are about half those quoted by McNeill (1980), this being due to the difference in definition of depth of investigation. For comparison with other electrical techniques, the figures quoted above are more useful (Barker 1989). Unfortunately, although horizontal coils allow greater depth of penetration, the system is more sensitive to errors of misalignment while at the same time the coils are more difficult to align. Measurements made with vertical coils generally appeared less noisy and more easily interpretable than profiles conducted with horizontal coils. All measurements were, therefore, made with vertical coils. In order to differentiate areas where the regolith attains thicknesses of more than 10 m, a maximum coil spacing of 40 m was normally employed with measurements being made along lines at intervals of 20 m.

The ground conductivity meter was used with three aims: firstly as an initial reconnaissance tool, to profile across wide valleys and plains in order to locate probable areas of thick overburden; secondly to investigate and locate precisely, linear features already observed on air photographs, such as joints, faults and dykes; thirdly, on a more local level where the overburden was thin, profiles were carried out in order to locate narrow fault and fracture zones not identified on aerial photographs in otherwise topographically featureless areas.

Normally the measured apparent conductivity was a weighted mean of the high conductivity of the regolith and the low conductivity of the bedrock. Therefore, in areas where the conductivities of the regolith and bedrock do not change, variations in the measured apparent conductivity are directly related to variations in the thickness of the regolith. Complications and ambiguities in the interpretation often arose where the conductivity of the regolith increases, generally towards the centre of a valley, as the clay content increases. The high apparent conductivities measured over near-surface concentrations of clay could then be wrongly interpreted as thick regolith. This ambiguity inherent in the interpretation of ground conductivity profiles, especially with measurements made with vertical coils, could normally be resolved by measurement of a single resistivity sounding on each broad anomaly.

Resistivity sounding

Quantitative results were obtained with the resistivity sounding technique, with rapid and accurate measurements being made with the offset sounding technique. This system of vertical electrical sounding enables the measurement of a conventional Wenner apparent resistivity curve but one in which the spurious effects of near-surface lateral resistivity variations are greatly reduced and often entirely eliminated (Barker 1981). The result is a smooth sounding curve which can be interpreted with considerable accuracy.

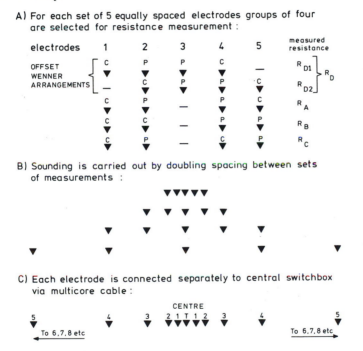

Fig. 3. Principles of Offset sounding.

The field technique employs a basic five-electrode array with measurements being made with five different electrode arrangements (Fig. 3). Two of these arrangements, D1 and D2, allow the measurement of conventional Wenner resistances, R_{D1} and R_{D2}, the average of the two providing the high quality offset Wenner resistance, R_D. In addition to providing further points on the apparent resistivity curve (Barker 1981), measurements at the other electrode arrangements, R_A, R_B and R_C, provide a check on the quality of the data through use of the tripotential relationship

$$R_A = R_B + R_C$$

which should hold for any geology and any electrode spacing (Carpenter & Habberjam 1956). An observational error defined as

$$e_o = \frac{(R_A - R_B - R_C)}{R_A} \times 100\%$$

is, therefore, a useful means of identifying instrumental problems.

The offset Wenner technique also provides a measure of the variation of the resistivity along the line of electrodes by computation of the offset difference, defined as

$$e_{off} = \frac{(R_{D1} - R_{D2})}{R_D} \times 100\%$$

For purely horizontal strata, e_{off} will be zero.

In this project, measurements were made with a Campus BGS-256 Offset Sounding multicore cable system and an ABEM SAS300 Digital Terrameter. At larger electrode spacings arrays of two or three steel stakes were implanted in the ground and a small amount of water was allowed to soak the soil around each electrode. This was found to be more than enough to combat the adverse effects of the dry surface soil in this area of Zimbabwe and to reduce the electrode contact resistance to an adequate level.

In the Victoria Drought Relief Project a total of 357 electrical resistivity soundings were measured at 52% of borehole sites. In order to process the considerable amount of data which quickly accumulates in a single day, all field measurements were stored on a microcomputer in the field office.

Data reliability

The quality of the field data was initially evaluated in the field where observational errors were calculated at each spacing. Where high errors, greater than about 4%, were recorded, the measurements were checked and corrected. The resulting high quality of the data is confirmed in Fig. 4 which shows the average r.m.s. errors for the 357 soundings calculated for each spacing. On average there is a slight increase in observational error with spacing, from 0.6% to 1.4%, reflecting the increased difficulty in measuring accurate resistances at larger spacings.

A single r.m.s. error computed for each sounding provides a means of estimating the quality of the sounding. A frequency distribution of the r.m.s. observed errors (Fig. 5) shows that in this investigation, soundings generally exhibited r.m.s. errors of less than 1%. r.m.s. errors of more than 3% were treated with caution. Fortunately, only seven soundings (2%) exhibited r.m.s. observed errors greater than 3% and only one greater than 4%.

The offset difference, e_{off}, was calculated at every spacing for every sounding. Although the offset difference can be influenced by the quality of the resistance measurement, it is generally a measure of the presence of lateral variations in earth resistivity. Over a truly horizontally layered earth the offset difference will be zero, but where depth to bedrock changes along the line of the electrodes or where the electrical properties of the regolith change laterally, the offset differences will attain high values.

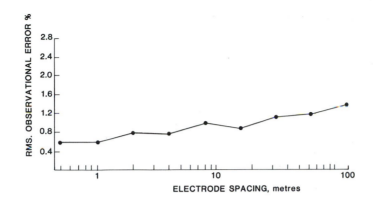

Fig. 4. r.m.s. observational error computed at each electrode spacing for 357 soundings in Victoria Province.

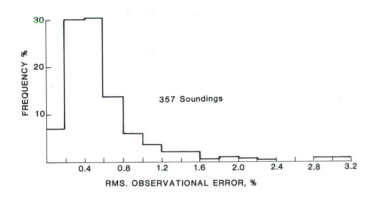

Fig. 5. Frequency distribution of r.m.s. observational errors computed for each of 357 soundings measured in Victoria Province.

r.m.s. average values have been computed at each spacing for the 357 soundings and are presented in Fig. 6. The smallest offset differences are normally found at a spacing of around 8 m but the offset differences increase at smaller spacings. This is likely to be the result of the greater error introduced at smaller spacings in the exact placing and depth of the electrodes. r.m.s. values increase at larger spacings where variation in depth to bedrock has an increasing effect.

A single r.m.s. offset difference was calculated for each sounding as a guide to the validity of using a horizontally layered interpretation. A frequency distribution of the r.m.s. values (Fig. 7) shows a normal range of 8–30% for this investigation. Only eight soundings (2%) exhibited r.m.s. offset differences greater than 50%, the interpretations from these soundings being treated with caution.

Theoretically, over a horizontally layered earth, a sounding curve should rise at no more than 45° (Fig. 8). An analysis of the 357 soundings measured in this investigation (Table 1) shows that almost 53% exhibited a final segment of curve rising with a gradient slightly steeper than 45°. This effect can be produced when poor electrode contact with the earth has resulted in the measurement of incorrect resistances, usually at larger electrode spacings. However, in every case where the field procedure was carefully checked by alternative field techniques and instrument settings, good electrode contacts and accurate resistance measurements were confirmed. It is certain, therefore, that the steep gradients result from strong lateral changes in earth resistivity.

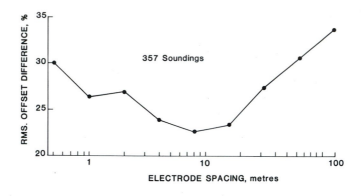

Fig. 6. r.m.s. Offset difference computed at each electrode spacing for 357 soundings in Victoria Province.

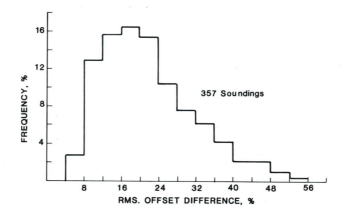

Fig. 7. Frequency distribution of r.m.s. offset differences computed for each of 357 soundings in Victoria Province.

Table 1. *Number of soundings exhibiting a steep gradient at large electrode spacings*

Electrode spacing at which gradient starts to rise at more than 45°	Number of soundings
4	0
8	0
16	22
32	71
64	95

Field studies also confirmed that gradients of more than 45° could often be expected. The normal field location of a sounding (Fig. 9) was in a basin or valley generally less than one kilometre across with kopjes on either side. In such a situation the centre of the sounding could be situated over relatively thick regolith whilst the outer electrodes would be located on thin regolith. In this case resistances measured at larger spacings become increasingly higher than would be expected over a constant thickness of regolith.

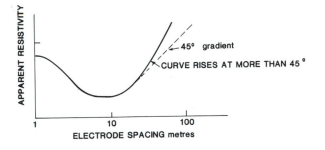

Fig. 8. Typical sounding curve with last branch rising at more than 45°.

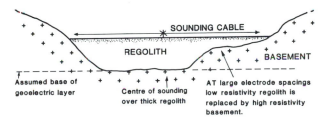

Fig. 9. Normal geological situation which could result in resistivity sounding curve of Fig. 8.

Interpretation

The initial interpretation of the apparent resistivity sounding curves was always carried out in the field so that the results could be used as a basis for an immediate borehole siting decision. In this project, interpretation was carried out using a simple partial curve matching technique as described for example by Koefoed (1979), the only necessary tools being a set of theoretical two-layer curves and one of generalized auxiliary curves. Interpretations conducted in this way were very accurate when, as was normally the case, a three-layer model resulted. However, the accuracy of this technique decreases with increasing number of layers involved. For this reason every interpretation was subsequently checked on the field office microcomputer using conventional techniques (Ghosh 1971). Only in a very few cases was it necessary, as a result, to revise the proposed borehole site.

The layered earth model obtained from a resistivity sounding interpretation is very much a simplification of the many different subsurface layers which may be present. As a general rule soundings were interpreted in the simplest model possible, this normally including three distinct electrical layers: high resistivity dry material overlying low resistivity water-saturated regolith overlying high resistivity unfractured bedrock.

Ambiguities in interpretation resulting from equivalence and suppression were often identified. Equivalence occurs when a subsurface layer has a resistivity which is lower (or higher) than the beds immediately above and below. In this case it is possible to define uniquely the product of the layer thickness and its conductivity but neither of these parameters independently. The water saturated zone on a sounding curve might be defined by a longitudinal conductance of 250 mS, but it would be impossible to determine whether this represented a 5 m thick layer with a conductivity of 50 mS/m (resistivity = 20 ohm m) or a 10 m thick layer with a conductivity of 25 mS/m (resistivity = 40 ohm m).

Problems of equivalence were generally much less significant than those caused by suppression. Suppression of a layer occurs where its resistivity is intermediate between the resistivities of the beds above and below. In this case the effect of the intermediate layer on the resistivity curve is only seen if the layer thickness is inordinately large. The main problem of suppression experienced in the Zimbabwe surveys was the difficulty in defining the layer of fractured bedrock which normally has a resistivity between that of the overlying saturated regolith and the high resistivity of the underlying unfractured bedrock. In these cases it was at least possible to use the field computer to estimate likely maximum and minimum thicknesses of potential fractured aquifer.

Hydrogeological interpretation

The role and application of geophysical techniques varied according to the field conditions. With experience it was possible to identify the following main types of hydrogeological situation: (1) valleys; (2) hills of outcropping granite basement, locally known as bornhardts and kopjes; (3) faults, fractures and joints; and (4) basic dykes. Detailed investigations of a number of representative field situations were carried out early in the siting programme to give insight into the problems likely to be encountered. Resistivity sounding was found to be particularly effective in the investigation of the first two of these situations. Examples of each are described below.

Valley: Gwana School

At Gwana School in Masvingo Communal Land a broad valley exists between two areas of outcropping granite. A profile of eight soundings was carried out across this valley in order to investigate the variation in thickness and resistivity of the regolith and bedrock. The interpretation is displayed as a geoelectrical section in Fig. 10 together with the results of a profile of ground conductivity measurements.

The regolith is clearly recognized and shows a number of important facets common to many basins encountered. Firstly, the resistivity increases consistently from 50 ohm m at its thickest development to 200 ohm m and above along the valley sides. This is interpreted as a clay-rich valley centre with sand-rich valley sides. Secondly, the thickest development of regolith is not coincident with the topographical low but is offset slightly to one side.

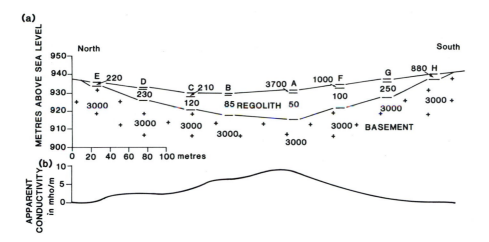

Fig. 10. (a) Geoelectrical section at Gwana School with resistivities shown in ohm m. (b) Conductivity profile measured with EM34 and 20 m vertical coil spacing.

In a broad valley of this type the thickening of regolith is not clearly associated with increased fracturing and successful borehole siting would rely heavily on the permeability of the regolith. A potential borehole site in this valley would, therefore, not be close to the topographical base but would have to be a compromise between the thickest development of regolith and a lateral move to avoid the impermeable clay-rich zone in the centre of the valley. Often the extent of the clay zone can be recognized in the field by a change in soil colour from grey to red and in the case of sharply defined clay zones the edge of the clay can also give rise to a line of springs.

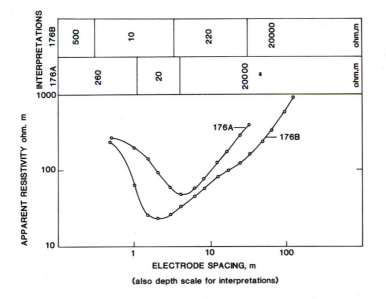

Fig. 11. Soundings measured at Salani Kraal. The slight change in slope seen on the rising branch of 176B represents the presence of a significant thickness of fractured and partially weathered basement.

Valley: Salani Kraal

Salani Kraal is situated in a broad valley where thick regolith was anticipated. No EM34 profile was carried out and a single sounding, 176A, was measured at the most practical proposed site as far as the community was concerned. This showed massive bedrock at only 4 m with an overlying clayey regolith. A second sounding, 176B, subsequently measured almost one kilometre away indicated a sandy regolith with up to 30 m depth to bedrock (Fig. 11). The difference in the shape of the resistivity curve brought about by the presence of fracturing in the basement is small and subtle. It is therefore important if potentially suppressed layers are to stand any chance of being identified that good quality field data be obtained. This is one reason why the offset sounding technique proved so useful in this project.

Here the thickening of regolith appeared to be associated with fracturing in the bedrock. A borehole drilled at the second sounding actually proved 23 m of gneiss regolith above fissured gneiss resulting in a good specific capacity of 0.18 l/s/m. The sharp and dramatic changes in thickness of regolith associated with basement fracturing, and demonstrated by this example from a gneissic area, were much less frequently encountered in granite areas.

Bornhardt: Mount Gondoi

Although a majority of boreholes were sited in valley situations, occasionally thick regolith and fracture zones were located close to hills or kopjes. Mount Gondoi is an excellent example of a granite bornhardt where concentrated runoff from the hill has developed a zone of deep weathering adjacent to it.

Conductivity profiles show high values for 200 m to the east and 100 m to the west coincident with the zone of deep weathering indicated by the resistivity sounding interpretation (Fig. 12). The variability of the resistivity of the regolith is substantial in this example. Particularly in the east it appears that clays have concentrated at the surface.

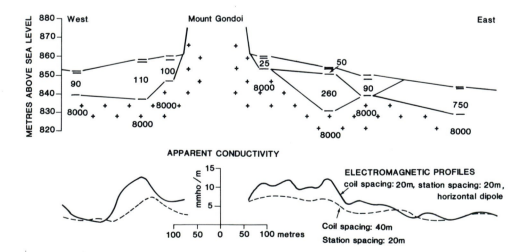

Fig. 12. Geoelectrical section and EM34 profiles across Mount Gondoi.

Potential borehole sites occur either side of the bornhardt where the regolith is thickest. In subsequent investigations in this type of situation one conductivity profile plus one sounding were normally sufficient to locate and confirm a borehole site.

Valley Head: Mhungu Kraal

Situations were often encountered in areas of gentle topography where the thickness of regolith appeared to increase towards the head of the valley. In the example shown at Mhungu Kraal (Fig. 13), the first sounding in the valley suggested only 11 m of regolith. However, a second sounding sited 400 m upstream showed 28 m of regolith with a slightly higher resistivity suggesting sandier conditions.

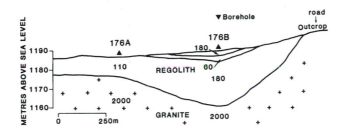

Fig. 13. Geoelectrical section at Mhungu Kraal.

Dykes and faults

In the investigation of subvertical features such as dykes and faults, resistivity sounding was of little use except for eliminating ambiguities caused by the presence of near surface clay patches. Profiling with the EM34 was generally successful in locating dykes and faults and any fracture zones which might be associated with them.

Discussion of interpretation

Formation resistivities

The values of regolith resistivity determined from the resistivity soundings show a typical range between 20 and 600 ohm m. The lower resistivities represent a clay-rich lithology while higher resistivities represent partially saturated or dry regolith. The range in regolith resistivity for the two major rock types of granite and gneiss shows considerable overlap (Fig. 14) although the mean resistivity for the granite regolith (236 ohm m) is higher than that for gneiss (145 ohm m). This difference reflects the greater ferromagnesian mineral content of the weathered gneiss.

Between the regolith and the unfractured bedrock, a layer of intermediate resistivity could often be identified. This was interpreted as fractured or partially weathered bedrock and was characterized by resistivities in the range 170–3000 ohm m. Unfractured bedrock generally has values above 3000 ohm m.

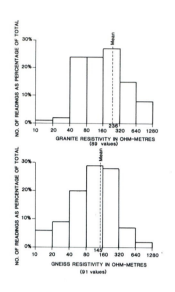

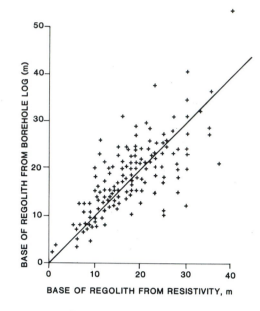

Fig. 14. Frequency distributions of regolith resistivity measured over (a) granite and (b) gneiss, with electrical sounding.

Fig. 15. Relationship between depth to bedrock observed in boreholes and that measured by resistivity sounding.

Depth to bedrock

The majority of resistivity soundings enabled determination of the depth to a high resistivity bedrock. This generally agreed within ±25% with that indicated by drilling (Fig. 15). Sources of error include inaccuracy in drillers' depth determination caused by the presence of corestones or irregular bedrock topography. A gradual change in weathering with depth was normally interpreted as a sharp vertical change in the resistivity, the depth to this often disagreeing with that determined in drilling.

Depth to water table

In nearly every case the depth to the top of the saturated zone estimated from the sounding interpretation was less than the standing water level observed in the boreholes. This was likely due to the presence of a thick partially saturated zone which can often be observed in hand dug wells overlying the saturated zone. This forms a layer of intermediate resistivity which often cannot be identified on the sounding curve.

Generally the resistivity determined depths were 2 m to 6 m shallower than the standing water level although discrepancies as large as 20 m were observed where the regolith consisted of clay. The data plotted in Fig. 16 include only data from regolith boreholes although similar results are obtained if all boreholes are used. Similar observations have since been made by Carruthers & Smith (1989).

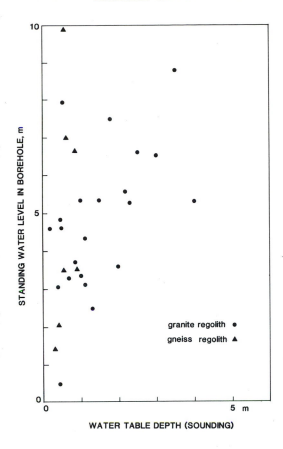

Fig. 16. Comparison of standing water level observed in boreholes and depth to water measured by resistivity.

Borehole yield

Shortly into the project, it became apparent that the highest yielding boreholes were associated with regolith resistivities which were neither too high nor too low. Very low regolith resistivities are associated with the presence of clay and low yields could be expected. The highest yielding boreholes are associated with regolith with resistivities between 100 and 600 ohm m. Below 200 ohm m the observed specific capacity falls as the clay content of the regolith increases. The yield also falls as the resistivity increases beyond 600 ohm m, this probably being due to the decreasing water saturation of the aquifer. The data used in Fig. 17 to demonstrate this relate to boreholes drilled only into the regolith.

Although the borehole yield could be related to the regolith resistivity, the latter was also observed to be related to the weathering history (White *et al.* 1988). However, this does not alter significantly the range of resistivity at which the highest yields could be expected.

Similar results are observed in gneissic areas although the resistivity range at which highest yields are observed is slightly lower at 150 ohm m–550 ohm m.

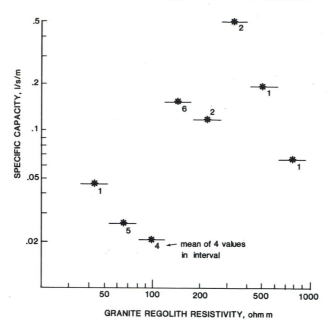

Fig. 17. Relationship between granite regolith resistivity and specific capacity of boreholes.

Borehole siting success rate

The fundamental objective in the borehole siting survey was the location of aquifers capable of sustaining a handpump discharge. The principal aquifers of the region are the regolith and the upper fractured zone of the bedrock.

Although a single fissure of less than 1 mm width could theoretically sustain a handpump yield, no known technique is available that can locate such a fissure. Even single fissures of many times this width are impossible to locate using surface geophysical techniques. Therefore, it is necessary to look for fissure zones—clusters of subhorizontal fissures near the top of the bedrock. It is generally assumed that their frequency is greatest where the regolith is thickest. Thus the rationale in borehole siting is to try to locate the deepest regolith on the assumption that this will maximize yields from both the regolith and the fissured rock. Furthermore, by seeking the greatest thickness of regolith, it can be presumed that the saturated thickness is greatest and therefore any borehole will maintain its yield for a longer period during any drought which lowers the water table.

In order to obtain an adequate yield it has been found that a minimum regolith thickness of approximately 10 m is required. This is in agreement with work by Omorinbola (1981, 1982) over Nigerian Basement rocks. Ideally, however, regolith thicknesses of more than 15 m are required, especially where it is known that groundwater levels are deep. Figure 18 shows the general tendency for specific capacity to increase with depth to bedrock for regolith aquifers. The wide scatter is probably due to different water levels. Resistivity is the only technique used which gives quantitative information on depth to bedrock although EM, if interpreted carefully, will give qualitative information. The results with resistivity are often so good that this technique could be used alone to locate zones of thick regolith.

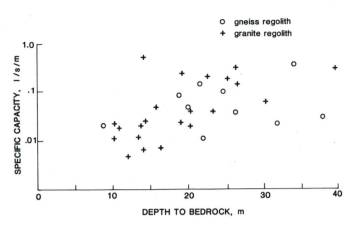

Fig. 18. Relationship between specific capacity and depth to bedrock for regolith aquifers.

However, when used alone it is a time-consuming search tool and so is better employed to complement the EM surveys. The EM surveys quickly eliminate the areas likely to be unsuitable and the resistivity provides a final quantitative check on aquifer thickness and formation resistivity.

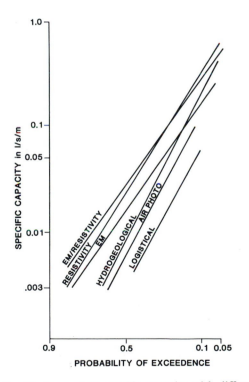

Fig. 19. Probability of achieving a given specific capacity with different siting techniques.

From the final calculations of specific capacity it was possible to construct the graph of Fig. 19 which shows the probability of achieving a given specific capacity for each of the siting techniques employed in the project. The results are summarized in Table 2 where the success rates for the various techniques are given regardless of aquifer type. These figures show that a great improvement in success rate is found when geophysical techniques are employed. The combination of EM and resistivity techniques proved to be better in this project than resistivity alone because in the time available for studying each proposed borehole site, a much greater area could be investigated.

Table 2. *Success rate for various siting techniques*

Technique	Percentage of successful boreholes
EM + resistivity	90
Resistivity	85
EM	82
Hydrogeology	66
Air photo	61
Logistical	50

Conclusions

The Victoria Drought Relief Project demonstrated the effective use of geophysical techniques, in particular resistivity sounding and electromagnetic surveys, in cost-efficient borehole siting. The detailed results presented here illustrate some of the problems which were encountered and the limitations of the techniques which were employed.

The vigorous archiving of all observed geophysical measurements during the project enabled detailed subsequent analysis and the important conclusions which have been drawn will be applicable to many similar basement areas of Africa.

The results of the Victoria Province Drought Relief Project are published by kind permission of the Ministry of Energy and Water Resources and Development, Zimbabwe, and the European Economic Community.

References

ACWORTH, R. I. 1987. The development of crystalline basement aquifers in a tropical environment. *Quarterly Journal of Engineering Geology*, **20**, 265–272.

BARKER, R. D. 1981. The Offset system of electrical resistivity sounding and its use with a multicore cable. *Geophysical Prospecting*, **29**, 128–143.

—— 1989. Depth of investigation of a generalized colinear 4-electrode array. *Geophysics*, **54**, 1031–1037.

CARPENTER, E. W. and HABBERJAM, G. M. 1956. A tripotential method of resistivity prospecting. *Geophysics*, **21**, 455–469.

CARRUTHERS, R. M. & SMITH, I. F. 1989. *Basement Aquifer Project 1984–1988: Summary report on surface geophysical studies*. British Geological Survey Technical Report WK/88/23.

EDWARDS, L. S. 1977. A modified pseudosection for resistivity and IP. *Geophysics*, **42**, 1020–1036.

GHOSH, D. P. 1971. The application of linear filter theory to the direct interpretation of geoelectric resistivity sounding measurements. *Geophysical Prospecting*, **19**, 192–217.

KOEFOED, O. 1979. *Geosounding Principles: Resistivity Sounding Measurements*. Elsevier, Holland.

McNEILL, J. D. 1980. *Electromagnetic Terrain Conductivity Measurement at Low Induction Numbers*. Geonics, Ontario.

OMORINBOLA, E. O. 1981. Components of saturated zone thickness in a Nigerian basement complex regolith. *Hydrological Sciences Bulletin*, **26**, 291–303.

—— 1982. Verification of some geohydrological implications of deep weathering on the basement complex of Nigeria. *Journal of Hydrology* **56**, 347–368.

WHITE, C. C., HOUSTON, J. F. T. & BARKER, R. D. 1988. The Victoria Privince Drought Relief Project, I. Geophysical siting of boreholes. *Ground Water*, **26**, 309–316.

From WRIGHT, E. P. & BURGESS, W. G. (eds), 1992, *Hydrogeology of Crystalline Basement Aquifers in Africa*
Geological Society Special Publication No 66, pp 203–220.

The use of ground electrical survey methods for siting water-supply boreholes in shallow crystalline basement terrains

R. M. Carruthers & I. F. Smith

British Geological Survey, Keyworth, Nottingham NG12 5GG, UK

Abstract. Resistivity methods are widely used as an aid to siting water-supply boreholes in basement terrain. Adverse conditions due to shallow bedrock can be identified readily but specific targets such as narrow conductive zones may be missed; quantitative interpretation may be less reliable than is often assumed. Electromagnetic (EM) profiling will detect localized zones of deeper weathering and the effects of fracturing at shallow depths, and it is used for the precise location of anomalous ground indicated by lineations on aerial photography. Routine EM surveys give better lateral resolution than resistivity for the conductive targets of interest. The 2-D and 3-D EM modelling techniques needed for interpreting the conductor geometry realistically are still under development.

Geophysical methods must be applied with due regard for the targets being sought and the local geological setting: they will not always be appropriate and resources should be targeted on districts where problems can be solved. Qualitative interpretations are adequate in some situations but more rigorous field procedures and analytical techniques are needed to ensure that useful hydrogeological information is obtained.

Large areas of crystalline basement in semi-arid tropical regions, more particularly in the southern parts of Africa and Asia, consist of granites and gneisses with a thin cover resulting from chemical weathering processes, largely controlled by the circulation of groundwater (Jones 1985). In a geophysical context the terms overburden, saprolite and weathered material tend to be used synonymously, rather than in any rigorous sense, when referring to this cover. As Palacky *et al.* (1981) describe for Burkina Faso in West Africa, the aquifer can be a three-component system: a perched water table which may disappear during the dry season; a zone of relatively high porosity and storage in the clay/quartz-rich residual products of advanced weathering (saprolite); a deeper zone of weathered and jointed rock containing fractures with high transmissivity but limited storage (saprock). The latter grades into fresh bedrock as the number of fractures diminishes.

The components of the weathered layer are generally discontinuous and an understanding of their relative contributions is important in determining the best exploration and development strategies: whether the aquifer should be exploited by shallow wells completed within the saprolite or by deeper boreholes which intersect productive fractures in the saprock. An unacceptable proportion of dry holes will often, though not invariably, be drilled if sites are selected only for the convenience of the intended users' community. Geophysical methods can help in identifying the more favourable locations for both dug wells and boreholes. Low yields (typically in the range 0.1–1 l/s) are utilized for domestic supply but in some areas it may be possible to find sites capable of providing for a larger population or for supplementary irrigation. Such sites are often associated with major fracture systems that are readily identifiable on satellite imagery or aerial photographs; geophysical surveys provide a means of fixing the precise location of these narrow features on the ground, so as to ensure that boreholes intersect them.

The results on which this paper is based were collected as part of a five-year research programme on the Hydrogeology of Basement Aquifers which was mainly funded by the Overseas Development Administration of the British Foreign Office; the fieldwork was undertaken in Malawi, Sri Lanka and Zimbabwe in collaboration with the local Water Departments.

Application of geophysical techniques

Resistivity sounding

Resistivity surveys have commonly been applied in the vertical electric sounding (VES) mode for estimating thicknesses and resistivities in a sub-horizontally layered sequence. In crude terms, the intermediate layer of lower resistivity in the typical sounding curve represents the target zone, lying between a more resistive cover and the bedrock. The parameters attributed to this zone are used to define suitable sites for a hand-pumped rural water supply, with recommended borehole siting criteria recognizing that there is a correlation between depth to bedrock, saprolite resistivity and borehole yield. Thus, in the 'Zimbabwe Master Plan', Martinelli (1984) specifies that for a successful drilled well the thickness of saturated overburden with a resistivity between 30 ohm m and 150 ohm m must exceed 25 m. The resistivity limits represent an attempt to quantify the constraints on permeability as set by increasing clay content at the lower level, and by the lack of sufficient weathering or fracturing within the bedrock at the upper level, though other workers have suggested that the upper limit can reasonably be extended to 400 ohm m (I. Clifford, Interconsult, Harare, pers. comm.). The minimum thickness requirement allows for the poor transmissivity that can be expected in this type of aquifer, as reflected in the low specific capacity of many boreholes. A high success rate is usually guaranteed if such recommendations are followed but the number of sites identified in convenient locations will be severely limited unless a thick overburden is common. Other targets and geophysical methods must be considered in districts where the weathered zone is known to be thinner, as in the Masvingo Province of Southern Zimbabwe.

In many cases, quantitative interpretation of VES results has proved to be unsatisfactory in this type of environment because it can give an inadequate or misleading impression. For instance, hydrogeological parameters such as effective porosity, specific capacity, permeability, yield, etc. cannot be predicted; the clay content of the saprolite and type of bedrock also remain ill-defined and the relationship with depths derived from boreholes is often poor. A degree of statistical correlation may be observed in large data sets (White et al. 1988), but the results at any one site rarely show much beyond a qualitative agreement, even for depths to bedrock.

A three-layer analysis is often sufficient to match the main features of a VES over shallow basement, with the curve typically showing a distinct minimum, as in Fig. 1: the effect due to the saprock is usually suppressed (see below) and thus the distinction between saprolite, saprock and fresh bedrock tends to reduce to one between a

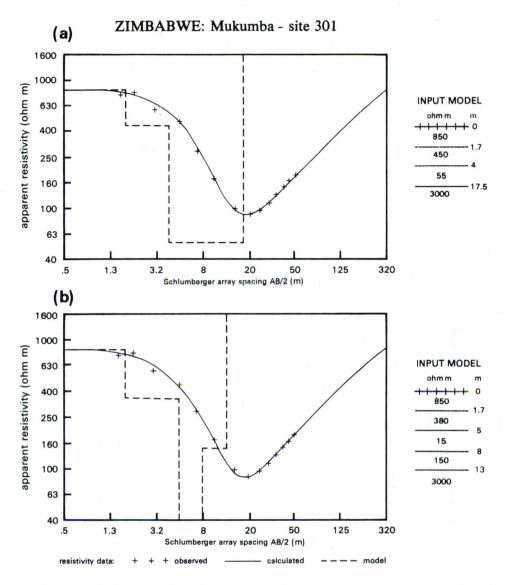

Fig. 1. Alternative interpretations of a vertical electric sounding showing suppression of a thin layer: (a) satisfactory fit achieved before drilling; (b) preferred model in the light of the borehole results.

weathered zone and bedrock. This is not to argue that more layers should not be inferred but that any routine VES interpretation on site—which is required practice in most production surveys—has to be treated with caution. While more complex interpretations are readily derived on portable computers which improve the fit with minor features of VES curves, their relevance will depend on factors such as borehole control, comparison with a number of adjacent soundings (to check for consistency in terms of a 1-D model), the experience of the interpreter in that area and the availability of a valid conceptual model, based on all the existing geophysical, geological and hydrogeological evidence. For these reasons, individual VES should not be interpreted in isolation if possible.

The main limitations to the method which have to be recognized (see Koefoed 1979 for more details) are:

1. *Equivalence*. A curve with a minimum or maximum derived from a three-layer case can be fitted within the accuracy of the field data by a range of 'equivalent' models in which the resistivity and thickness of the second layer are varied significantly; this range can be surprisingly wide given a large resistivity contrast with the adjacent layers, and the values attributable to the saprolite can often be adjusted by a factor of five or more without the quality of the curve match being adversely affected.

2. *Suppression*. The influence of layers which are thin in relation to their depth tends to be too subtle for their presence to be recognized in VES curves. The addition of thin layers with high resistivity contrasts into a model will reduce the interpreted depth to bedrock, while layers of intermediate resistivity can increase it. Borehole control may indicate that the models have been over-simplified: Fig. 1 illustrates a case where it was necessary to include a more conductive saprolite above a layer attributed to fractured, hard rock in order to match the driller's depth of 8 m to bedrock. Suppressed layers of this type are thought to be a factor in many cases.

3. *Lateral variations*. Lateral changes in either layer thickness or resistivity significantly increase the uncertainty when fitting a 1-D model to the data. While the problems due to inhomogeneity in the near-surface layer are often apparent as irregularities in the VES curve, the distortions can be more subtle. Comparison of a number of soundings, and especially orthogonally-oriented pairs of soundings, provides evidence of the degree of lateral variation. Figure 2 illustrates how the presence of lateral effects can be detected in this way. An east–west VES expansion sub-parallel to a boundary produced a flatter curve, while the north–south VES from the same centre indicated the presence of less resistive ground. Traversing along the latter direction confirmed that the conductance of the saprolite reduced markedly across a contact which lay about 20 m to the south of the site. It should be noted that there is no obvious discontinuity in either curve which would suggest any marked lateral change. Electrical anisotropy, as might exist where thin graphitic bands occur, can produce similar effects.

An effect noted in a large proportion of VES from Zimbabwe was that the final ascending limb rose more steeply than the theoretical maximum which would result from a substrate of very high resistivity beneath a conductive overburden: when plotted in the standard way, with equal logarithmic x–y axis scaling, the limiting angle for a Schlumberger array is 45° (as in Fig. 1b). The point in the curve at which the distortion becomes significant is often unclear, leading to errors in the parameters interpreted for the overlying layers; this is usually more important than uncertainty in estimates of the true resistivity of the bedrock. No fully satisfactory explanation has been found for the scale on which this occurred. The effect was more common in areas of high contact impedance and may result partly from systematic errors in measuring low-voltage signals. Errors due to coupling or the use of a low-frequency, alternating current source should be negligible although similar problems have been associated specifically with multicore cable systems. Lateral variations in conductance and relief on the basement surface appear insufficient as an explanation: 2-D model studies could not fully reproduce the effect; estimates of the magnitude of the uniform dip required (Koefoed 1979) are also incompatible with layer thicknesses and outcrop distribution.

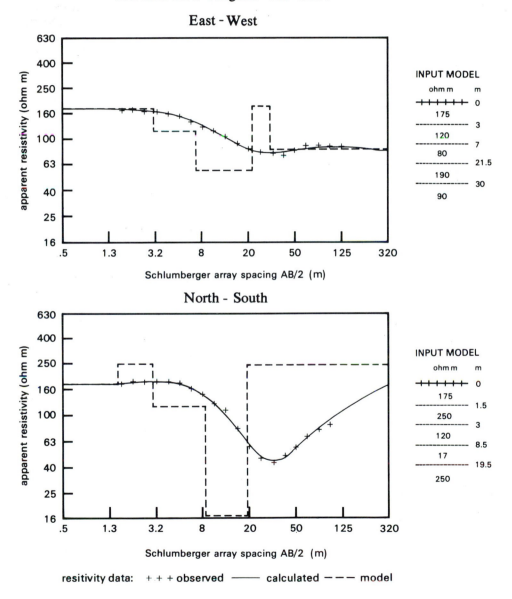

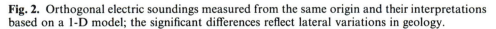

Fig. 2. Orthogonal electric soundings measured from the same origin and their interpretations based on a 1-D model; the significant differences reflect lateral variations in geology.

The interpretations of VES need to be related explicitly to the aquifer and its properties, as far as this is possible. There tends to be a tacit assumption in the literature that the top of the low resistivity layer represents the water table; in fact, a closer examination of case histories, combined with practical experience, often shows this not to be the case. A moist zone, as seen in shallow dug wells, can be associated with the base of the uppermost (high-resistivity) layer comprising desiccated soil, leached saprolite or laterite. This moisture may represent perched or pendular, infiltrating water closely associated with clay particles and retained by surface tension. However, it does not bear an obvious relationship to the effective water table, or to a capillary fringe, even though the latter may be over a metre thick: Smith and Raines (1988) show an example from Chinembire, Zimbabwe where there is about 25 m between the top of the low resistivity layer and the water table as intersected in a borehole.

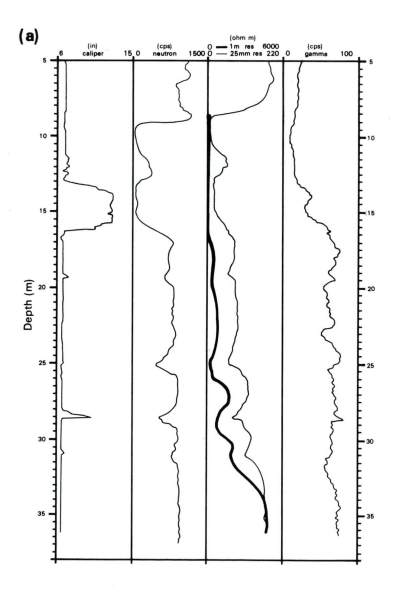

Where there is little or no control on the type of model which should be applied within a given area it may be useful to consider the total longitudinal conductance of the sequence overlying resistive bedrock. A value for this can be derived directly from the steeply-rising limb of a sounding curve (Orellana & Mooney 1966) and checked against the value obtained by summing the contributions (of thickness divided by resistivity) of the individual layers in an interpreted model. The total conductance is defined within much closer limits than any layer interpretation and so it may give a more reliable comparison of conditions between different sites. Clearly, being an aggregated quantity, the total conductance contains less information on the nature of the sequence and it still needs to be considered in relation to observed apparent resistivities and likely limits on layer thicknesses.

Borehole logs can provide more direct evidence of the variations in physical properties with depth, for comparison with surface measurements and lithological control. Figure 3 shows a suite of logs obtained from a high-yielding well in Zimbabwe in which a short-term pump test gave a specific capacity of 0.76 l/s/m with a yield exceeding 2 l/s; the VES data are from the same site. There is a general increase in resistivity with depth although values for the 1 m normal log only exceed 3000 ohm m below 32.5 m whereas values of greater than 5000 ohm m are more

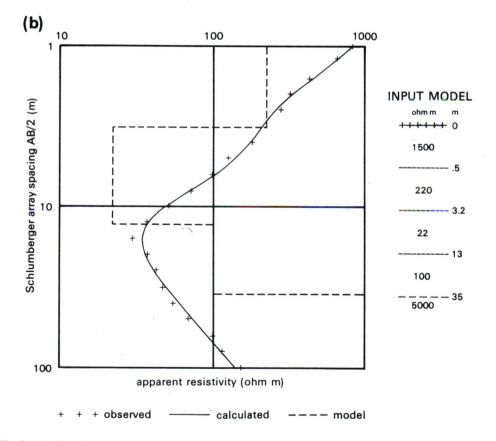

Fig. 3. (a) Geophysical logging and (b) surface resistivity (VES) data from a borehole site at Rungai, Zimbabwe.

typical of the bedrock in this district; the driller's log noted 'harder rock' at 11 m, corresponding with the top of a less porous, resistive band. The interpretation given for the VES was based partly on the borehole information and while it indicates a degree of consistency between them this is more apparent than real: it illustrates that a satisfactory model can be derived but a wide range of alternatives is also available because of the uncertainties described above. In particular, the section of the VES associated with the fourth layer (i.e. near and beyond the curve minimum) shows irregularities and it could be attributed to near-surface inhomogeneity distorting a simpler four-layer curve, to give a much shallower depth to fresh bedrock. The borehole logs also imply that the base of aggressive weathering is near 18 m, while the coincident local excursions in resistivity, neutron porosity and caliper response (e.g. near 28 m) are indicative of the individual fractures (Shedlock 1990) which probably provide major contributions to the borehole yield. Such details cannot be derived from the VES data.

The use of 2-D or 3-D models shows the effects of more complex structures on resistivity soundings. Thus, it can be demonstrated that there is little possibility of a water-bearing fracture at depth producing a discrete discontinuity in a VES curve (Carruthers 1988): for such 'breaks' to occur at an electrode spacing related to the depth to the aquifer (Ballukraya *et al.* 1983), the features would have to be sub-horizontal rather than near-vertical and induce some non-linearity into the relation between current flow and potential field. The occurrence of irregular, narrow conductive pathways along fractures and joints, combined with some clay filling, might give variable current gathering effects which are not allowed for in standard theory, but a true causal relation between fracturing and 'breaks' has yet to be established. In general, VES are inappropriate for detecting localized fracture systems.

Resistivity traversing

Simple resistivity traversing techniques have been used with some success to identify and trace localized anomalous zones, almost invariably those showing a lower resistivity. Data collected using an array with a fixed electrode separation may show significant anomalies providing that the array is sufficiently sensitive to changes within the depth range of interest, but the origin of any anomalies detected will be ambiguous: the current electrode separation effectively controls the depth of investigation while the potential electrode separation and station interval determine the lateral resolution.

It is possible to quantify changes in the depth to bedrock for a simple two-layer situation by covering the same ground with a pair of separations, selected on the evidence of resistivity soundings: the closer spacing gives the resistivity of the upper layer alone while the wider spacing defines the total longitudinal conductance above the resistive substrate.

The type of anomaly attributable to a single, narrow water-bearing fracture will not usually be detectable above the background of geological 'noise', but the effect of a broader zone of fracturing within the bedrock should give a resolvable anomaly. A larger response may be expected where the fracturing is reflected in a local thickening of the overlying saprolite due to enhanced weathering, or where the conductivity is increased due to saline water, clays, pyrite or graphite contained within the fractures.

More sophisticated field and data processing procedures can significantly improve the definition of anomalous zones but they are not readily applied to rapid reconnaissance surveys.

The microprocessor-controlled resistivity traversing system (MRT) (Griffiths & Turnbull 1985) has been developed to provide coverage which is a compromise between profiling and full VES, the data being interpretable in the form of a depth section through the ground. This method was tested with some success in Zimbabwe (Griffiths 1987), indicating that quite detailed information can be obtained, as shown in Fig. 4 for a traverse crossing a lineament. However, the method is still relatively slow in comparison with electromagnetic (EM) systems for routine field use in identifying fracture zones.

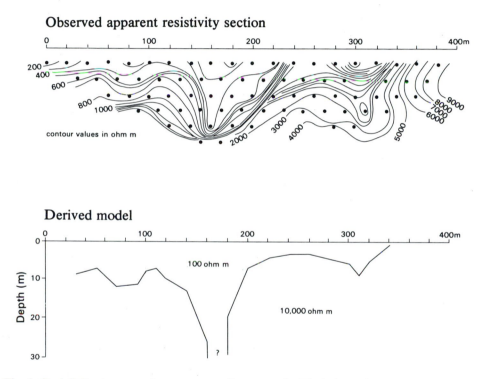

Fig. 4. Resistivity traverse data across a lineament in Zimbabwe shown above a 2-layer interpretation (after Griffiths 1987).

Electromagnetic (EM) conductivity mapping

The role of resistivity methods in traversing is being superseded by the increasing availability of EM instruments, which are faster and require fewer people for the fieldwork. The EM methods also have two technical advantages in semi-arid basement terrain.

(i) inductive coupling between the transmitter, receiver and the ground avoids the need for direct electrical contact, thus eliminating problems associated with resistive, hard, dry or stony surface conditions;

(ii) a better combination of depth of investigation and lateral resolution of anomalies is achieved for a given coil separation (as compared to the length of a resistivity array), improving the chances of identifying narrow conductive bodies.

Electromagnetic induction has been widely applied in geophysical surveying for minerals using a range of transmitter and receiver configurations. Systems based on the moving-source principle, with two portable coils linked by a reference cable, have become important tools in groundwater exploration more especially since their sensitivity has been increased and instrumental drift minimized. The output can be given directly in the form of apparent ground conductivity as in the Geonics EM34 instrument (McNeill 1980) so that the data can be assessed easily in the field. It is important to remember, however, that the readings only approximate the true bulk conductivity of the ground and as values increase the divergence becomes greater. In addition, the instruments are insensitive to resistive bodies: if the target has a resistivity greater than about 150 ohm m, resistivity traversing methods will provide better discrimination and should be used in preference, despite their reduced convenience (Smith & Carruthers 1990). EM techniques are most appropriate when looking for fracture zones or pockets of thicker saprolite in regions with relatively shallow depths to bedrock (<20 m).

Variations in apparent conductivity measured using a single coil configuration may indicate changes either in the thickness or conductivity of a layer: the ambiguity cannot be resolved without additional information. However, the depth of investigation is influenced both by the coil separation (set to 10 m, 20 m or 40 m with the EM34) or coil orientation (vertical or horizontal), giving a variety of combinations in practice: when the coils are vertical and co-planar this depth is significantly less than in the horizontal mode, making the instrument more sensitive to the upper, relatively conductive, overburden. (It should be noted that orientations are sometimes defined in terms of the dipole moment; this lies along the axis of the coil and so the terms vertical and horizontal become interchanged.) Figure 5 is an example of EM traverse data across a dambo (a characteristic type of shallow valley found in Malawi) which illustrates the change in response as the coil configuration varies.

Table 1. *Predicted EM34 values for the models illustrated in Fig. 1 (Makumba site 301)*

	EM coil spacing (m)	apparent conductivity (mS/m)	
		vertical coils	horizontal coils
Resistivity model (a)			
850 ohm m to 1.7 m	10	6.5	9.6
450 ohm m to 4.0 m	20	7.8	8.2
55 ohm m to 17.5 m	40	7.0	4.6
3000 ohm m			
Resistivity model (b)			
850 ohm m to 1.7 m	10	9.5	13.6
380 ohm m to 5.0 m	20	10.5	9.2
15 ohm m to 8.0 m	40	8.3	3.9
150 ohm m to 13.0 m			
3000 ohm m			

Depth interpretations can be made either using inversion techniques or by computing the EM response expected from a given model. The latter technique can be useful for reducing the ambiguity when modelling resistivity soundings as EM equivalence operates within a different range. An example of this is given in Table 1 which shows the predicted EM34 values for the models illustrated in Fig. 1.

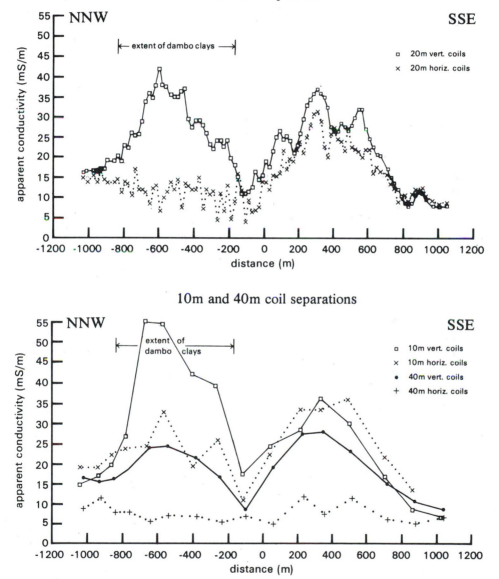

Fig. 5. Ground conductivity traverse data, illustrating the different responses to a thin cover of highly conductive, smectitic clays (to the NNW) and thicker, moderately conductive saprolite (to the SSE).

The EM values measured in the field differed somewhat from those modelled, as might be expected with a variable cover of thin saprolite, but the ratios of values from different coil configurations can be used to characterize the models. The observed ratio given by 10 m : 20 m horizontal coil values was about 1.6, compared with 1.2 from model (a) and 1.5 from model (b); the observed ratio for vertical : horizontal coil values at 20 m separation was 1.15–1.2 as against 0.95 for model (a) and 1.14 for model (b). Thus, the EM data clearly suggest that model (b) is more likely to be correct, as proved in the drilling.

ZIMBABWE: Chikore School EM traverse no.4

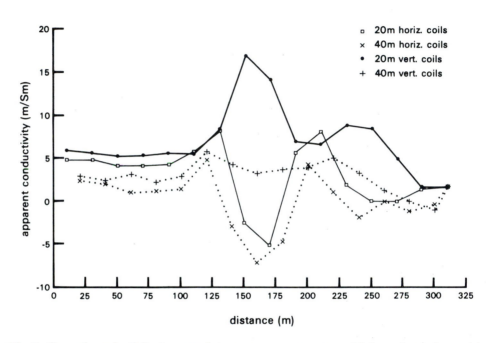

Fig. 6. Ground conductivity traverse data across a narrow steep-sided conductor located by the minimum in the horizontal coil response.

While the EM methods are effective in conductivity mapping, their particular advantage over resistivity traversing is in the detection of relatively narrow, dipping conductors. The EM anomaly given by a thin, steeply dipping conductor (often referred to as 'dyke-like') comprises two apparent conductivity maxima on either side of a minimum, centred approximately over the conductor. It is important to note that the conductivity maxima form part of a single, composite anomaly and they occur over the resistive ground flanking the conductor; they must be distinguished from real increases in bulk conductivity within the underlying ground. The response is best seen in the horizontal co-planar coil data (as in Fig. 6) because of the more effective coupling between the primary EM field and conductor for this configuration, combined with its greater depth of penetration. Some information on the conductance and attitude of the body can be derived from this type of anomaly using standard techniques (for example Telford *et al.* 1976) although, as discussed below, care must be taken in the application of these methods. The width of the anomaly between the maxima as measured at the background level should equal the coil separation plus the width of the conductor; the amplitude of the minimum is related to the conductance of the target.

This type of anomaly proved to be of considerable interest in Masvingo Province, Zimbabwe (Smith & Raines 1987) where extensive lineaments identified from aerial photographs had been correlated with the surface expression of fracture systems beneath thin saprolite (Greenbaum 1986, 1990). Traverses were run across the lineaments, which often coincide with vleis (the local equivalent of the dambos found in Malawi), using an EM34 with coil separations of 20 m and 40 m in both horizontal and vertical modes. Figure 6 is a representative example of the response obtained with the vertical coil configuration profile showing a simple maximum corresponding to the compound anomaly from the horizontal coils. No examples of minima with flanking maxima were obtained with the coils vertical, although with the deeper penetration afforded by the 40 m vertical coils there was occasionally a suggestion of this pattern developing, superimposed on the relative high values. In areas of thicker overburden the horizontal coil response also reverts to showing the maximum, characteristic of the vertical coils, which reflects only the local increase in conductance: Beeson & Jones (1988) describe a technique for locating relatively narrow water-bearing zones based on drilling conductivity highs, taking a ratio of horizontal to vertical coil readings of greater than 1 in combination with VES results as an indication of deep weathering.

These EM anomalies are clearly associated with lineaments and provide significantly greater precision than aerial photography alone for positioning a borehole to intersect any related fracture system. However, there are reasons to suppose that the anomalies result predominantly from steep-sided bodies of limited depth extent, rather than from the conventional semi-infinite, parallel-sided 'dyke' model. An interpretation based on the latter model indicates the presence of bodies with widths of up to 20 m and conductances (conductivity–width products) of up to 2 mho. Two geological models consistent with drilling evidence might be proposed in explanation:

(1) very conductive material (clay or groundwater) contained in a set of relatively thin fractures extending to a significant depth, separated by solid rock;
(2) moderately conductive material across the full width of the anomalous zone and confined essentially within the saprolite.

The groundwater is usually fresh, with conductivities typically in the range 10–100 ms/m (Shedlock 1987), and it is unlikely that the necessary conductance could be achieved based on model 1, as recognized by Lindqvist (1987). Thus, model 2 is preferred, with the fractures encouraging groundwater circulation and preferential weathering in their upper parts to produce thicker saprolite locally, as shown in Palacky et al. (1981). The depth to which the conductor is developed in model 2 would be limited by geochemical processes and the extent of groundwater circulation, in the same way as saprolite development is controlled elsewhere.

The dip of a conductor can be inferred simply from asymmetry in the shape of the EM anomaly and it can be quantified on the basis of a 'dyke' model within a resistive medium: the body dips beneath the greater (in both amplitude and area beneath the curve) of the two flanking maxima. Dip is clearly an important factor in ensuring that a borehole intersects the zone of interest beneath the water table. However, such an interpretation assumes that the conductor is thin, extensive both along strike and with depth, and is parallel-sided: all these criteria will rarely be met if model 2 applies. More sophisticated modelling, using both numerical and analogue techniques, is required to define the relative importance of changes confined essentially to the weathered material, as against the effect of a fracture zone within the bedrock itself; the techniques to enable this to be done using either approach are currently under development.

Appropriate EM field procedures must be adopted when narrow conductive zones are potential targets:

(i) horizontal coil data are essential;
(ii) a number of traverses are needed across a structure to confirm continuity and sufficient strike extent in the anomaly pattern;
(iii) a station interval should be established which is small enough to define the target properly;
(iv) 10 m coil data should be collected if possible as these can give useful information on the nature of the saprolite and identify anomalies due to conductive superficial clays.

The depth of investigation is related to the coil separation and the conductivity layering within the ground: guidelines which suggest this depth is about 1.5 × (coil spacing) with the coils horizontal apply only to the limits of detectability for a conductive layer in otherwise uniform ground; in practice, a thin vertical target will need to be much shallower than this.

Other EM techniques

The (very low frequency) VLF–EM method has also been used successfully to map fracture zones in this type of environment. It has the advantages of being very portable and simple to use by one operator, and gives a good response to large-scale features of only moderate conductivity. However, its reliance (for most practical purposes) on existing transmitters restricts the strike of the features that can be detected, while the depth of investigation is limited where conductive overburden occurs. Tests using VLF methods in both Malawi and Zimbabwe proved unsuccessful because the received signal strengths from transmitters in Europe, United States of America and Australia were too low in that part of Africa.

The most promising recent developments in EM techniques have been in the pulsed or Time-Domain EM systems (TEM), as compared to frequency-domain systems such as the EM34. It is now possible to obtain TEM soundings which give information from within 10 m of the surface, bringing them into the range needed for shallow basement terrain studies. Their main advantage over VES lies in the much reduced volume of ground sampled for a given depth of investigation; this makes them less susceptible to the errors caused by the averaging out of lateral variations, and means that bedrock topography can be mapped in greater detail. The traversing and sounding modes are also more efficiently integrated than for either resistivity or frequency-domain EM methods. A combined interpretation of resistivity and TEM sounding data will generally lead to a significant reduction in the problems of equivalence which occur when either is used on its own. EM34 data can serve this purpose to an extent but the constraints are less well defined due to the limited number of depth sample points and uncertainty in the zero reference level. However, the effectiveness of TEM equipment has yet to be proved in this field environment and it may suffer from low signal strengths in the more resistive conditions.

Survey considerations

Electrical survey methods are particularly effective for rejecting potential borehole sites, where no adequate conductive target can be identified. Thus, traversing can be used to eliminate large tracts of unproductive ground when looking for fracture zones, while VES will usually provide evidence of shallow bedrock even if they cannot be interpreted quantitatively for assessing the overburden aquifer. However, the characterization of the more marginal targets is not straightforward. A combination of geophysical methods looking at different properties of the ground will usually allow for a more complete, reliable interpretation and may be justified in difficult areas or when looking specifically for high yielding wells. Seismic, radon 'sniffing' and magnetic surveys can be used to complement the electrical techniques if the necessary time and resources are available.

The amount of geophysical survey and the quality of its interpretation in the context of siting rural water supply wells is severely restricted by economic factors, with contracts typically requiring sites to be identified at a rate of two to three each day. In these circumstances it is unlikely that high success rates will arise solely from the use of geophysical surveys unless very specific targets exist. Knowledge based on previous work in similar hydrogeological environments, together with an intelligent inspection of the ground, can in themselves provide an effective way of siting wells. Our experience indicates that answers to some of the important specific questions raised during development programmes cannot be provided on a routine basis by the geophysical field techniques and interpretation methods currently employed; nor is there any evidence that the rule-of-thumb interpretations applied by some local contractors are of themselves any more successful. The attributes and limitations of the methods and the interpretations must therefore be recognized by the contracting agency and project manager as well as by the geophysicist if an effective survey procedure is to be established.

The nature of the aquifer determines the geophysical method which is most appropriate in each case and it is necessary to have a clear concept of its type, thickness, depth and permeability. While we have distinguished between the sub-horizontal overburden aquifer and the sub-vertical fracture type, the two will often be related. In the former case, we must be aware that a major factor in obtaining adequate borehole yield could be the presence of a thin, but highly fractured transition zone at the top of the saprock; in the latter case, it will not be possible to identify the individual, productive fractures within a conductive zone. The geophysical logging of boreholes can provide this type of information and its use may be justified on a limited basis as a means of controlling the interpretation of surface methods. Survey procedures may need to be adapted in the light of results obtained to ensure that the anomalies are identified and defined in sufficient detail, and the amount of time which can be spent at any one site should also be flexible for the same reason.

The concept of success rate and its relation to the amount of geophysics undertaken is widely discussed but is often very misleading unless the full range of contributory effects are considered. For instance, in an area where the aquifer is well understood, geophysics may not significantly affect the already high success rate. In new or difficult areas, where the success rate is based on few results or is very low, the application of carefully chosen methods could greatly improve the results; alternatively, localized success may be lost in the assessment for a region as a whole. Reliance on a simple yield criterion can be misleading; it is important to consider depth to water, well recovery and drawdown effects and long-term yield (because minor structures might give a high initial yield which decreases significantly with time) in relation to borehole performance. In summary, it is necessary to use geophysics flexibly, bearing in mind the relative importance of individual sites and overall survey costs, rather than by rote.

Conclusions

Geophysical techniques make a valuable contribution to well siting procedures, given that there will be considerable variation in their application between different tropical regions of the world related to climate and geology: the results described have been mainly derived from areas in southern Africa where depths to both bedrock and the water table were shallow. In order to make a significant improvement to the efficiency with which successful sites are identified the methods should be chosen to suit the geological environment, on the basis of available information, and economic constraints. Specific points to consider are as follows.

1. The emphasis in areas of shallow bedrock should be on profiling to locate fracture systems inferred from aerial photography. When using moving-source frequency domain EM equipment, it is essential that horizontal coil data are collected in sufficient detail.

2. Where the saprolite constitutes the most effectively exploitable resource and is more than 20 m thick, then VES should be applied.

3. Where the resources of equipment or expertise are limited, geophysical surveys should be concentrated in new areas and any districts where failure rates are known to be high; acceptable success rates might be maintained elsewhere using surface features and experience of the area.

4. The application of geophysical surveys should not be constrained unduly by contractual specifications. Survey parameters such as EM coil orientation and spacing, station interval, line length and separation, number of VES measured and their maximum electrode separation, must be adapted to site conditions.

5. Geophysics should not be undertaken in isolation but fully integrated with the results of geological, hydrogeological and drilling activities.

6. Previous interpretations should be reviewed continuously in the light of drilling results to establish:

(a) areas of greater or lesser difficulty;

(b) areas with specific targets (for example: deep water table, dykes, poor quality water etc.);

(c) hydrogeophysical models which can be used to reduce the ambiguities in interpretation.

7. An effective database, allowing efficient archiving and retrieval, can be a considerable help in reviewing and assessing existing data.

8. Where high yields are required a range of methods (such as seismic refraction and magnetics combined with resistivity and EM) is probably justified to provide the complementary data which will increase confidence in the overall interpretation.

9. Further studies needed to enhance the application of geophysical methods in this environment include:

(a) field trials with time-domain EM equipment;

(b) 3-D analogue and mathematical modelling of the horizontal loop EM response;

(c) compilation and correlation of reliable drilling data, including geophysical logging, with surface geophysical results;

(d) relating the detectability of fracture systems to overburden thickness.

This paper is published with the permission of the Director, British Geological Survey. The work on which the paper is based was supported by the Overseas Development Administration and was undertaken in collaboration with the Ministry of Energy, Water Resources and Development in Zimbabwe; the Public Works Department in Malawi; the Water Resources Board in Sri Lanka.

References

BALLUKRAYA, P. N., SAKTHIVADIVEL, R. & BARATAN, R. 1983. Breaks in resistivity sounding curves as indicators of hard rock aquifers. *Nordic Hydrology*, **14**, 33–40.

BEESON, S. & JONES, C. R. C. 1988. The combined EMT/VES geophysical method for siting boreholes. *Ground Water*, **26**, 54–63.

CARRUTHERS, R. M. 1988. *Basement Aquifer Project: Geophysical Studies to Evaluate Groundwater Resources in Crystalline Bedrock Terrains in Malawi, Sri Lanka and Zimbabwe, 1986–1988*. British Geological Survey, Technical report, WK/88/10.

GREENBAUM, D. 1986. *Comments on 1986 Field Structural Investigations in Masvingo Province, Zimbabwe and Notes on Follow-up Geophysical Surveys*. Unpublished report of British Geological Survey, Overseas Division.

—— 1990. Lineament studies in south-east Zimbabwe. *In: Proceedings of Geohydrology in Africa—Workshop on Groundwater Exploration and Development in Crystalline Basement Aquifers, Harare, Zimbabwe, 1987*. Commonwealth Science Council Technical Paper 273, series number CSC(89)WMR-13.

GRIFFITHS, D. H. 1987. *Mapping Weathered Basement Rocks in Zimbabwe using a Microprocessor Controlled Resistivity Traversing System: a Progress Report*. Unpublished report of University of Birmingham, Department of Earth Sciences.

—— & TURNBULL, J. 1985. A multi-electrode array for resistivity surveying. *First Break*, **3**, 16–20.

JONES, M. J. 1985. The weathered zone aquifers of the basement complex areas of Africa. *Quarterly Journal of Engineering Geology*, **18**, 35–46.

KOEFOED, O. 1979 *Geosounding Principles 1. Resistivity Sounding Measurements*. Elsevier, Amsterdam.

LINDQVIST, J. G. 1987. Use of electromagnetic techniques for groundwater exploration in Africa. *Geophysics*, **53** (3), 456–458.

MARTINELLI, E. 1984. National masterplan for rural water supply and sanitation. Report of Ministry of Energy and Water Resources, Government of Zimbabwe.

McNEILL, J. D. 1980. *Electromagnetic Terrain Conductivity Measurement at Low Induction Numbers*. Geonics Ltd. Technical note TN-6.

ORELLANA, E. & MOONEY, H. M. 1966. *Master Tables and Curves for Vertical Electrical Sounding over Layered Structures*. Madrid, Interciencia.

PALACKY, G. J., RITSEMA, I. L. & DE JONG S. J. 1981. Electromagnetic prospecting for groundwater in Precambrian terrains in the Republic of Upper Volta. *Geophysical Prospecting*, **29**, 932–955.

SHEDLOCK, S. L. 1987. *Borehole logging in Masvingo Province, Zimbabwe*. Unpublished report of British Geological Survey, WD/OS/87/7.

SMITH, I. F. & CARRUTHERS, R. M. 1990. Application of surface geophysical methods to exploration for groundwater in crystalline basement rocks: the key issues. *In: Proceedings of Geohydrology in Africa—Workshop on Groundwater Exploration and Development in Crystalline Basement Aquifers, Harare, Zimbabwe, 1987*. Commonwealth Science Council Technical Paper 273; series number CSC(89)WMR-13.

—— & RAINES, M. G. 1987. *Geophysical Studies on the Basement Aquifer in Masvingo Province, Zimbabwe*. Unpublished report of British Geological Survey, Regional Geophysics Research Group, Project Note 87/7.

—— & —— 1988. *Further Geophysical Studies on the Basement Aquifer in Masvingo Province, Zimbabwe*. Unpublished report of British Geological Survey, Regional Geophysics Research Group, Project Note 88/4.

TELFORD, W. M., GELDART, L. P., SHERIFF, R. E. & KEYS, D. A. 1976. *Applied Geophysics*. Cambridge University Press.

WHITE, C. C., HOUSTON, J. F. T. & BARKER, R. D. 1988. The Victoria Province Drought Relief Project, I. Geophysical siting of boreholes. *Ground Water*, **26**(3), 309–316.

From WRIGHT, E. P. & BURGESS, W. G. (eds), 1992, *Hydrogeology of Crystalline Basement Aquifers in Africa*
Geological Society Special Publication No 66, pp 221–242.

New approaches to pumping test interpretation for dug wells constructed on hard rock aquifers

R. Herbert, J. A. Barker & R. Kitching

British Geological Survey, Maclean Building, Crowmarsh Gifford, Wallingford OX10 8BB, UK

Abstract. Large-diameter hand-dug wells are used to exploit the weathered zone aquifer of hard rock areas. A pumping test is described which is suitable for such dug wells. This test can be diagnostic of aquifer type and hydraulic properties.

Four techniques are described which can be used to interpret the pumping test. The first technique uses a digital model to simulate aquifer conditions and to reproduce changes in well water level observed during the test. Multi-layered conditions and partial penetration of the dug well can be simulated providing radial symmetry pertains. Two techniques are then described which require only manual calculation. In both these cases the aquifer is assumed to be single-layered but in one the aquifer is assumed to be confined and in the second it is unconfined. The fourth technique uses software developed for use on a personal computer. The software simulates a leaky aquifer and automatically determines the hydraulic parameters of the two layers of the leaky system using a least-squares technique. Finally, a week-long test is described which can be used to predict empirically the long-term behaviour of both dug and collector wells.

1. Introduction

Millions of large-diameter dug wells exist in the hard rock areas of the world. Whilst many techniques for dug well testing have been described in the literature (Barker & Herbert 1989), none was found entirely appropriate to the particular nature of the weathered zone aquifer of hard rock areas. Because of this the British Geological Survey have developed new methods of pumping test interpretation which suit their understanding of this complex aquifer.

Such tests and their interpretations are useful in that they allow estimates to be made of the local aquifer properties. Such knowledge may give insight into why some zones of weathered hard rock are more productive than others. Also knowledge of such properties and observation of well performace can allow predictions to be made of the likely long-term sustainable yields obtainable from dug wells.

2. The weathered zone aquifer

The weathered zone of basement areas is complex and variable. Wright *et al.* (1988) give a full description of factors affecting the make-up of this complex aquifer and the reasons for its variability. It is shown that geolgly, lithology, weathering history, mineralogy, tectonics and geomorphology can all affect the details of geometry and hydraulic properties of the weathered aquifer.

Figure 1 shows all the components of a typical weathered profile in a hard rock area. Where rainfall is high enough (>200 mm/annum) the water table is usually within the clayey saprolite. Most of the water movement in this profile is horizontal in the brecciated top of the saprock lying just above fresh bedrock.

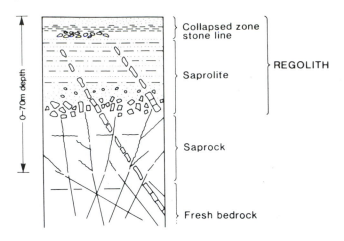

Fig. 1. Idealized weathered profile above crystalline basement rocks. (1) Collapsed zone. This may show marked lateral variations being generally sandy on watershed areas with illuviated clay near the base and sometimes a stone line changing to predominantly neoformed clay minerals in valley bottomlands (dambos). Slope bottom laterites may also occur associated with the peripheral dambo clays. (2) Saprolite is derived by in situ weathering but is disaggregated. Permeability and effective porosity tend to decrease at higher levels as a consequence of increase in secondary kaolinite minerals. (3) Saprock is cohesive weathered bedrock.

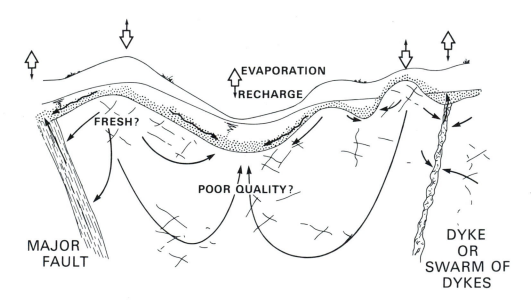

Fig. 2. Stylized groundwater movement on a regional scale.

Figure 2 shows groundwater movement on a regional scale. The weathered aquifer is connected to the fractured fresh rock aquifer below but in general the major components of flow are along the base of the weathered zone, which is fed mostly by vertical recharge from above and also discharges most of its flow by evaporation from topographic lows.

Figure 3 is based on Holmes (1966) and illustrates how rugose the interface between fresh and weathered rock can be. It is worth quoting the reference above, which although now somewhat simplistic still illustrates graphically the development and nature of this interface. 'Falconer pictured a region of granite and gneiss rotted by chemical weathering to considerable but unequal depths according to the composition and structure of the various rocks ... It is well known that certain granites suffer deep sub-soil decomposition where closely spaced joints facilitate the downward migration of groundwater'. Figure 3 illustrates different phases in the development of such weathered zones. The weathered zone can vary in thickness markedly from location to location. In much of Sri Lanka the interface seems particularly rugose. Outcrops of 'relict' boulders regularly appear 10 or 20 m from 10 m deep, successful dug wells. Wright *et al.* (1988) identify similar situations in Malawi and Zimbabwe, but at St Nicholas in Zimbabwe, deep weathered zones of a more uniform thickness have also been identified.

(a) Vertical section of fresh granite with varied spacing of joints.

(b) After a period of rotting by percolatry ground-water down to the level of permanent saturation, decomposed rock, black.

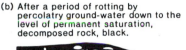

(c) Decomposed rock partially removed leaving a variable thickness to basement.

Fig. 3. Stages in the evolution of weathered basement.

From the view of pumping test interpretation, probably the most important thing to recognize is that the weathered zone aquifer can be highly non-homogeneous: it is multi-layered and, accordingly, no one single set of aquifer parameters can be expected to describe fully a dug well's performance. Thus, the task of the pumping test analyst is formidable.

2.1. The idealized aquifer section

Figure 1 shows the generalized section assumed to exist at each dug well site, for the purposes of this paper. If the water table is within the clayey saprolite, flow in the system can be simplified as occurring in two interconnected layers, the clayey saprolite and a lower more permeable layer comprising the saprock and lower brecciated saprolite lying above fresh bedrock. Figure 2 shows that in reality there can be a component of flow from the fresh bedrock upwards to the dug well but this has been assumed to be negligible for the majority of dug well locations. The aquifer system is not the same at every site. If the clayey layer is relatively impermeable compared to the lower more permeable layer, then water will flow horizontally towards the dug well within the confined lower layer. In contrast, if the permeability of the clayey layer is only about 20 to 200 times less than that of the more permeable layer, then significant leakage will occur, fed by the water table, vertically through the clayey zone into the predominantly horizontal flow of the lower layer. This is the classic leaky aquifer situation (Hantush 1956). As the permeability of the clayey layer gets nearer to that of the lower layer, flow through it will no longer be vertical, but flow in the two layers will combine together to form a complex three-dimensional pattern. Finally, when the clayey layer has the same permeability as the lower layer, the aquifer behaviour will conform to that of the classic unconfined aquifer (Boulton 1954).

All the above aquifer situations have been met during the BGS programme of dug well pumping tests.

3. Five pumping test techniques for interpretation of tests on dug wells

3.1. Introduction

A simple field procedure has been developed for a recovery test on a dug well which is appropriate for practical regular use in the field and which can be analysed directly by any of four analytical procedures described below. This field procedure is described fully elsewhere (Barker & Herbert 1989) and is also summarized in §3.2.

Finally, an empirical field test is described, which was designed to assist in the prediction of likely long-term (six-monthly) safe yields of collector and dug wells. This kind of test should be carried out in the field, wherever such predictions are of importance.

3.2. The field procedure for a dug well recovery test

The well geometry must be determined. Figure 4 shows the nomenclature used for interpreting the test.

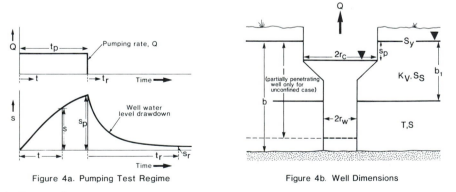

Figure 4a. Pumping Test Regime   Figure 4b. Well Dimensions

Fig. 4. Dug well recovery test, notation for pumping test regime and well and aquifer dimensions.

The well should be pumped for time t_p, at a reasonably constant rate, which must be measured. The drawdown must be monitored, and at intervals (of say 10 minutes initially) the quantity λ (see Equation (1)) is calculated. The test should continue until the value of λ falls to 0.95, and preferably lower.

If the aquifer is unconfined, and the drawdown exceeds about 10% of the saturated depth of the well before a suitably low λ value is reached, it may be necessary to repeat the test, after recovery, using a lower abstraction rate. Experience in a given aquifer should soon indicate a generally suitable rate of abstraction for this type of test.

The maximum drawdown, s_p, at the end of pumping, should be noted. Then as a minimum requirement the times, t_r, taken for the well to recover by 25%, 50% and 75% should be measured. Also, if the tester has access to use of the analytical model, described in §3.3.3., the full hydrograph of the recovery should be recorded, manually or automatically, as this will allow more accurate application of that particular technique.

The quantity λ is the ratio of water taken from storage in the well to the total volume of water pumped at the time t_p:

$$\lambda = \pi r_c^2 s_p / Q t_p \tag{1}$$

where r_c is the internal radius of the dug well casing, s_p is the drawdown in the well at the end of pumping, Q is the rate of pumping and t_p is the period of pumping. Any set of consistent units may be used in Equation (1).

3.3.   *Techniques to analyse the field dug well recovery test*

Four techniques are described for interpreting recovery tests on dug wells. Firstly, a finite difference model is described, which requires a large fast mainframe computer to run it and which has been used to carry out detailed generalized studies of dug well behaviour and for detailed studies of particular dug well tests, like those of the two examples given in the next section. It is not practical for this kind of model to be used for the routine analysis of dug well tests.

Secondly, two techniques are described which simulate the extremes of the likely boundary conditions for the aquifer system encountered. They are easy to apply in the field and easy to interpret using hand calculators and graphical methods. Both these techniques have been published previously: (Herbert & Kitching 1981; Barker & Herbert 1989). It is these two techniques which were designed for routine application to dug well pumping tests.

Thirdly, an analytically exact model is described, which can simulate large diameter well tests in a leaky or confined aquifer. Any leakage is assumed to be vertical in the upper layer. A water table and its associated relatively high release from storage can also be simulated as the upper boundary to flow in the leaky layer. This method would only be suitable for use by those with access to a relatively sophisticated personal computer (preferably with maths co-processor). It does, however, allow rapid analysis of some important aquifer boundary conditions. For example, its use could replace the graphical interpretation of Barker & Herbert (1989), which is suitable for confined aquifers. In addition it could simulate the type of leaky aquifer condition where vertical leakage occurs, i.e. where the permeability of the upper clayey aquifer is more than about 20 times less than that in the lower more permeabile zone of the aquifer system.

3.3.1. The finite difference model. The model is two-dimensional and simulates flow to large-diameter wells in aquifer systems having radial symmetry. It is a finite difference model with solution by a successive overrelaxation technique incorporating a predictor subroutine. A typical nodal network uses a 16 (vertical) by 26 (radial) mesh. The vertical node spacings can be varied according to the aquifer layering present. The radial node spacings are increased with the radius so that improved resolution is obtained in the region near the well, where the greatest variations take place. The distant radial boundary condition is 'no-flow' and the radial nodes are arranged so that this boundary is sufficiently far from the well for no significant drawdown to occur there. Horizontal and vertical permeabilities can be varied independently of each other over all nodes. It is necessary to specify the specific yield at the water table and the specific storage of the whole profile. A typical initial time step is five minutes, increasing throughout a stress period. A change of pumping rate (e.g. cessation) requires a new stress period and reversion to the initial time step. The model is run on a Cray IS computer at University of London Computer Centre.

Unlike all the other techniques described below, this model can simulate flow in more than two layers, it can represent a moving table and flow through aquitards can be inclined to the vertical. This model is therefore capable of representing a greater variety of real aquifer conditions more closely than any of the other techniques. As such it has been used occasionally to simulate dug well pumping tests where results from the more simple field tests are ambiguous. Two such examples of its use are presented in §4.

In addition the model has been used to study the characteristics of dug well behaviour in a general way. Wright *et al.* (1988) describe the results of those tests. Various generalized results will be mentioned below when appropriate.

3.3.2. Two analytical techniques appropriate for use in field. Two methods of analysis have been developed which assume either of the possible extremes of aquifer boundary conditions pertain, i.e. the aquifer is homogeneous and unconfined or the aquifer is totally confined within the saprock by the clayey layer. The tests have been described fully elsewhere (Herbert & Kitching 1981; Barker & Herbert 1989), are simple to carry out in the field and do not require computers or sophisticated mathematics for their interpretation.

3.3.2.1. Analysis of the confined aquifer condition. Barker & Herbert (1989) describe the test in full. This method of analysis assumes the aquifer is fully confined by the clayey layer, i.e. K_V of Fig. 4 is zero. As described above, the well geometry is known, the well has been pumped for a short period, t_p, and the times t_r, for the well water level to recover by 25, 50 and 75% have been monitored. Nomograms have been developed, which relate the dimensionless numbers, ρ, (t_r/t_p) to τ, $(4\pi T s_p/Q)$ and ρ to a, (Sr_w^2/r_c^2) for a range of values of λ. These nomograms have been developed for 25, 50 and 75% recoveries of well water level. Values for ρ and α are known directly from observation hence θ and a can be obtained from which T and S, the hydraulic parameters of the aquifer, can be obtained directly. Figure 5 shows the nomograms deduced for the 25% recovery condition.

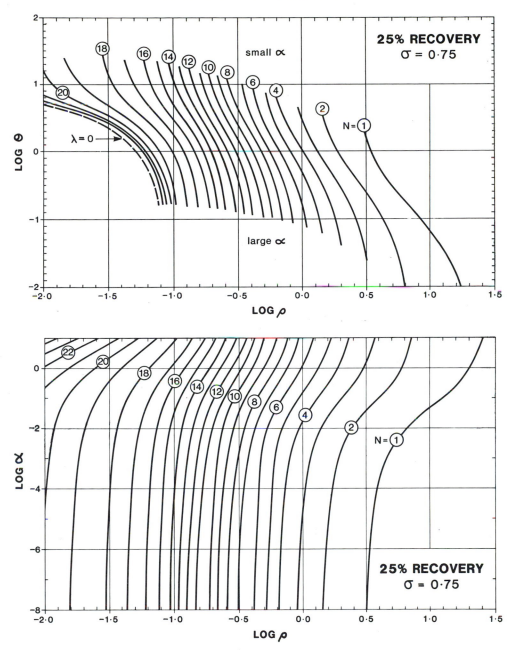

Fig. 5. Nomograms for interpretation of dug well recovery test on confined aquifers.

The results of applying the confined-test analysis to leaky situations have been studied. If the aquifer is homogeneous and confined, i.e. the clayey regolith is virtually impermeable when compared to the saprock, the test will give accurate values for the transmissivity and storage coefficient of the lower confined layer. On the other hand if the clayey layer can transmit a significant amount of water relative to that transmitted by the lower, more permeable, layer, the response in the early stages of the recovery will be something like the response of a leaky aquifer. Also, a homogeneous non-layered, unconfined aquifer can be considered as an extreme case of this 'leaky' condition.

The method used was to simulate tests using the finite-difference model (§3.3.1) and then to analyse the resulting drawdown curves using the confined-test analysis. The aquifer simulated had a total transmissivity of $10 \, m^2$/day and T_1 is the transmissivity of the lower 3 m of a 10 m thick aquifer. S_y was 0.05 and S_c (the sum of both layers) was 0.001. The results are summarized in Table 1.

Table 1. *Effect of layering on results of recovery test*

	T (m²/d) Derived			S (−)			Notes
T_1 modelled (m²/d)	25	50	75	25	50	75*	
3	10^{-6}	2	7	10^6	0.2	0.02	Water table aquifer
7.5	10^{-3}	2	7	10^3	0.1	0.01	Layered—leaky
9 and 9.5	0.05	2	8	6	0.02	0.003	Layered—leaky/confined

* Percentage recovery

If results for T and S obtained in the field have the same pattern as those in Table 1, a 'leaky' situation is indicated and only the 75% recovery results are indicative of the aquifer situation. In fact, about half of all the results on dug wells carried out to date display the kind of behaviour noted above.

3.3.2.2. *Analysis of the unconfined aquifer condition.* Herbert & Kitching (1981) describe this test in full. The test requires that 50% recovery of well water levels is achieved after a period of pumping. In fact, the original paper described two methods which could be used to interpret the pumping test. One of the methods required 90% recovery to occur. In practice this is hardly ever achieved in a reasonably short time so it is ignored here.

The 50% recovery test assumes the aquifer is homogeneous and unconfined. For this method of analysis, the well can be partially penetrating, l is the full depth of the well below the water table and b is the full saturated thickness of the aquifer. Also, $r_c = r_w$ is assumed. Most of the water is eventually derived from a lowering of the water table. The pumping test is carried out in the field as described in §3.2 but in addition to determining the well geometry an estimate of the total thickness of aquifer, b, of Fig. 4, must be obtained. For this test only one value of T is estimated, that is for the 50% recovery condition. No estimate can be made of the storage parameters. The transmissibility of the unconfined aquifer, T, is calculated using Equation (2), by substituting values for all the other parameters observed during the pumping test.

$$T = [a/(F.b'^{(c + d)})]^{1.0/(c + d - 1.0)} \qquad (2)$$

where $a = t_r/r_w^2$, $F = 0.104 \, (b/r_w^T)^{1.02}$, $b' = t_p/r_w^2$, $c = -0.127 \ln(b/r_w^T)$, $d = -0.332 \ln(l/b)$ and r_w^T is given by: $r_w^T = b(K_V/K_H)^{1.2}$ where K_V/K_H is an assumed anisotropy for the otherwise homogeneous, single layered, unconfined aquifer.

The same field procedure is used as for the confined test and the geometry is also defined by Fig. 4. There is a 95% chance that T will lie within one half to twice the value calculated. Equation (2) was derived empirically from statistical analyses of a large number of simulations of pumping tests using the finite difference model described earlier. Equation (2) is easily evaluated using a programmable hand-held calculator.

3.3.3. An analytical model for leaky and confined aquifers. Software has been developed for use on a personal or mainframe computer for simulating and analysing recovery tests in large diameter wells in a semi-confined aquifer (Barker 1989). This software is still under development in terms of improving user-friendliness. Two programs can be run. The first accepts data of pumping test behaviour and then by automatic variation of aquifer parameter values finds a best fit for any specified set of aquifer parameters required. Usually T, S, K_v, S_s and S_y are chosen to be determined. The second program accepts a set of values for T, S, K_v, S_s, b_1, r_w, r_c, Q, t_p and S_y and will then print out an analytically exact hydrograph appropriate to the pumping test for that set of parameters.

There are some restrictions on the applicability of the software. It is assumed the water table does not move, therefore only confined aquifers with small drawdowns should be analysed using this method. The well must be fully penetrating and most importantly it is assumed that leakage in the semi-confining layer is vertical, i.e. the answers will be meaningless if K_v of the semi-confining layer is not less than 1/20 of the equivalent permeability of the main aquifer. An inspection of the best fit results obtained from the programmes will normally indicate whether the parameter values determined are reasonable or not, and hence whether the assumptions of the analytical solution are valid.

The programs have been installed on an IBM PC compatible battery-powered portable, which has a maths co-processor card installed. It can therefore be used in the field at the pumping test site. Each parameter fitting takes about five minutes and each hydrograph print out takes three minutes or so.

It should be noted here that most real aquifers for dug well locations are extremely non-homogeneous and there is no real distinction between the confining layer and the lower more permeable layer, but that these gradually meld into each other. Hence, we cannot expect perfect fits from application of this software.

3.4. An empirical long-term field test to predict safe yield

This test was devised primarily to give an estimate of the long-term yield of a collector well during normal irrigation use. To date this test has only been used by BGS on collector wells, which are dug wells with horizontal adits. However, the test is just as useful for dug wells.

The well is pumped for three two-hour periods each day for about ten days. The discharge is selected so that the well will not run dry during this period. The drawdowns of the well water level are monitored at the start and end of each pumping period. Experience shows the drawdown is in two parts. There is a roughly constant daily change in drawdown from the start to end of pumping each day, this is s_{OE}, see Fig. 6. Also, there is a gradual fall in the background level of drawdown, s_0, which is recorded just before start of pumping every day. Most importantly, if s_0 is plotted against the log of time passed since start of the test, the plot is linear. This plot can therefore be extrapolated to predict the background drawdown that would occur after a complete dry season. Drawdowns are roughly proportional to discharge and therefore knowing the head in the well available for drawdown it is possible by simple proportion to predict the safe discharge of the well that can be maintained over one dry season.

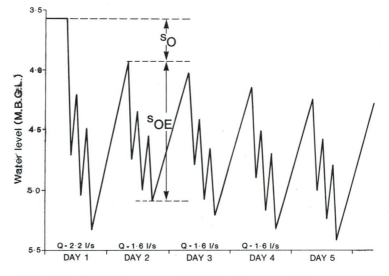

Fig. 6. Typical record of drawdowns recorded during long-term field test at each change of pumping rate.

Thus, if the background drawdown (obtained by graphical extrapolation) at the pumping test rate, Q_0, after 100 days is s_{0100} and the extra drawdown, which occurs at the end of every third pumping period, is s_{0E} (a constant), and if the allowable drawdown in the well is s_A, the safe discharge Q_S is given by:

$$Q_S = Q_0 s_A / (s_{0100} + s_{0E}) \tag{3}$$

This value is of course only approximate. If near-by hydrogeological boundaries occur the value can be altered significantly.

In addition to the above, the standard Jacob method is applied to the test using the straight line portion of the semi-log plot in the usual way and a value of 'regional' transmissivity is obtained. Experience shows that often this value is significantly different from that obtained from the dug well tests. This is probably due to the varied nature of the regolith.

4. Examples of application of BGS methods

4.1. *Mponela, North Malawi*

Wright *et al.* (1988; Appendices), give a full description of the dug well related work carried out at Mponela, in Dowa District, Central Region, Malawi. The work included geological surveys, site investigation boreholes, a geophysical survey, construction of a large-diameter dug well which was later converted to a collector well having three radials with a total length of 88 m, and water chemical testing. In addition there was a full pumping test programme carried out and it is some aspects of this which are now re-studied in a little more detail.

A simplified lithology in the dug well was as follows:

    0–2 m   clays
    2–4 m   silt, clayey
    4–7.5 m   silt to coarse sands, clayey
    7.5–11 m   saprock
    11–12 m   saprock, harder, more stoney

To translate the above to the idealized section of Fig. 1, there was a clayey confining layer down to 7.5 m then a more permeable aquifer down to 11 or 12 m (3.5 to 4.5 m thick), where the relatively impermeable bedrock was met. The 2 m diameter dug well was therefore fully penetrating. There were strong indications from geophysics and drilling that depth to bedrock was markedly variable within tens of metres of the dug well site.

4.1.1. The small drawdown dug well pumping test, 26/11/86. The details of the field pumping test are given in Table 2 and the results determined assuming the confined aquifer condition are summarized in Table 3.

Table 2. *Details of small drawdown recovery test at Mponela North*

Level of datum above ground level	1.015 m
Static water level below datum	5.12 m
Saturated aquifer thickness	7.895 m
Duration of pumping, t_p	100 mins
Pumping rate	0.4 l/sec
Well water level at t_p	5.644 m
Well drawdown, s_p	0.524 m
Time to 25% recovery	42 mins
Time to 50% recovery	105 mins
Time to 75% recovery	155 mins

Table 3. *Application of confined aquifer nomograms to small drawdown test at Mponela North*

	Transmissivity deduced, T, (m²/d)		
% Recovery	25	50	75
	22	27	very large
	Storage coefficient deduced, S (–)		
% Recovery	25	50	75
	4×10^{-5}	5×10^{-5}	very small

Applying the unconfined aquifer condition for 50% recovery and assuming $l = b = 8$ m and $K_H/K_V = 1$ gave a value for T of 12.4 m²/d. Read §3.3.2.2.

For analysis using the analytical model for leaky aquifers, the following constants were used in all cases: $r_c = 1.0$ m; $r_w = 1.0$ m; $Q = 34.56$ m^3/d; $t_p = 0.069$ days.

The times and drawdowns used in the pumped well for the optimization procedure for best fit selection of T(m^2/d), $S(-)$, K_v(m/d), S_s(m^{-1}), $S_y(-)$ and b_1(m) are given in Table 4. A good fit to the data could not be achieved when K_v was set to 0 so it was assumed the aquifer is 'leaky'.

Table 4. *Times and drawdowns used in the optimization procedure for the pumped wells*

% Recovery	Time (days)	Drawdown (m)
0	0.069	0.524
25	0.098	0.393
50	0.1419	0.262
75	0.176	0.131

Further trials using the model with $K_v > 0$ suggested that S_y did not affect the accuracy of the fit. This means the field test was not long enough to draw down the water table sufficiently for S_y to have an effect on the test results. So in subsequent fitting trials S_y was kept constant at an arbitrary value.

In the final trials it was ascertained that none of S, K_v and S_s could be determined with any accuracy but that T had a preferred value of 16 m^2/d having 50% confidence limits of lying within 12 to 39 m^2/d.

The best fit values found by the analytical model for leaky aquifers were as follows: T(m^2/d) = 16 (50% confidence limits of 12 to 39); S, K_v and S_s are significant but unknown with any certainty; S_y does not contribute to the test fitting.

4.1.2.   A pumping test on a slim borehole 20 m from Mponela North. A slim borehole, MP1, was drilled to bedrock 20 m from the dug well site. This was pumped at 1.5 l/s for five hours and its recovery was monitored. The plot of residual drawdown versus log of (total time since start of pumping/recovery time) was linear and using Jacob's recovery analysis (Todd 1959), gave a transmissivity for that location of about 7.8 m^2/day.

4.1.3.   The finite difference model used to simulate the large drawdown tests carried out on the dug well and the collector well (see §3.3.1). In using this model a best fit for the aquifer parameters is found by making a succession of manual changes to each parameter in turn followed by a commonsense selection of the optimum path the subsequent changes should follow to best reproduce the pumping test hydrograph obtained in the field. Small drawdowns cannot easily be modelled by this technique so only a series of deep drawdown tests carried out on the dug well and the subsequent collector well tests were simulated. These tests are described in full by Wright *et al.* (1988).

During modelling it was thought that better correlations were obtained when the model was layered and the results assumed, rather arbitrarily, that 75% of the total transmissivity occurred in the bottom 3 m. The specific yield, S_y, was selected as 0.5 and the storage coefficient, summed over both layers, was assumed to be 0.001. The transmissivity results were as follows:

Dug well deep drawdown 24–30 m²/d;
Collector well deep drawdown 28 m²/d;
Collector well long-term test 28 m²/d.

Figure 7 shows the F.D. model's hydrograph prediction for the deep drawdown on the collector well compared with the observed in the field test.

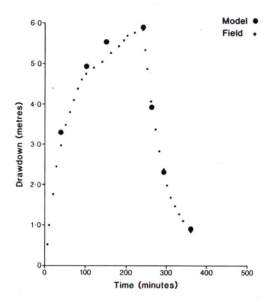

Fig. 7. Modelling of a deep drawdown test run on the collector well at Mponela, 22 July 1987.

4.1.4. Conclusions on the results from the various short-term tests. The large range of results and the odd anomalous result show how difficult it is to make accurate interpretations of pumping tests on dug wells in the regolith. Ideally, the application of the two analytical techniques described in §3.3.2 should be enough to draw satisfactory conclusions. In this case that could not be done. What can be said with some certainty is that the aquifer is to some degree leaky. The F.D. model, the analytical model and to some extent the application of the confined condition of Barker & Herbert (1989) (§3.3.2.1) all indicate this. Thus, we can probably reject the Herbert & Kitching (1981) result of §3.3.2.2 in this instance as being inappropriate. Also, remembering that the aquifer is non-homogeneous we can say the aquifer is probably leaky having a transmissivity in the range 12 to 39 m²/day and a specific yield of 0.1 or greater.

4.1.5. The long-term yield test at Mponela North. The details of a long-term test carried out on the collector well follow. The dug well at Mponela was converted into a collector well by drilling out three adits to a total length of 88 m from the main shaft. The following data were recorded when pumping tests were carried out.

Static water level: 4.648 m below datum (1.25 m agl)

Pumping regime: Daily 0.600–08.00; 11.00–1300 and 16.00–18.00 modi-
                            fied during first and last days on account of delayed start
                            but maintained six hours daily pumping nevertheless
                            (Table 5).
Pumping rate: 4.0 litres/sec
Well water levels: See Table 5 and Fig. 8.

Table 5. *Well water levels during long-term test, Mponela North collector well*

	Levels (m) at start and end of each pumping phase					
	Times of reading levels					
Date	09.30	11.30	12.45	14.45	16.00	18.00
30 August	1.609	4.830	2.421	5.122	2.557	5.336
	Adjusted Times					
	06.00	08.00	11.00	13.00	16.00	18.00
31 August	1.712	5.156	1.956	5.264	2.003	5.100
1 September	1.767	5.284	2.011	5.322	2.164	5.365
2 September	1.814	5.253	2.033	5.329	2.097	5.384
3 September	1.848	5.228	2.062	5.373	2.128	5.352
4 September	1.876	5.271	2.092	5.372	2.160	5.343
5 September	1.904	5.330	2.122	5.451	2.193	5.497
	Adjusted Times					
	06.45	08.45	11.45	13.45	16.45	18.45
6 September	1.929*	5.368	2.145	5.416	2.210	5.471
7 September	1.957 at 06.45 hours. End of test.					

* 1.934 m at 06.00 hours

During the long-term test water levels were also taken in borehole MP1, 6 m
distant from the easterly directed radial. These levels were taken 10 minutes before
the start and 10 minutes before the end of each day's pumping and are plotted in
Fig. 9.

Semi-log plots were made of the drawdowns at the beginning of the first daily
pumping cycle in the collector well and observation well (Fig. 10). The apparent
transmissivity by the Jacob analysis is in the range 58–60 m^2/day.

4.1.5.1. Long-term yield for one dry season. The drawdown at 180 days by
extrapolation of the semi-log plot is 0.71 m, s_{0180}, with additional drawdowns of
3.45 m, s_{OE}, for a 4 l/s pumping rate. The total available drawdown to the pump
suction is 6.9 m, s_A. Thus, using Equation (3):

$$\text{Maximum pumping rate} = 4 \times 6.9/(0.71 + 3.45) = 6.6 \text{ l/s.}$$

Thus the intermittent pumping rate which will be consistent with the total
available drawdown will be about 6.6 l/s.

This long-term prediction assumes that no adverse boundary conditions come into
play during the course of pumping.

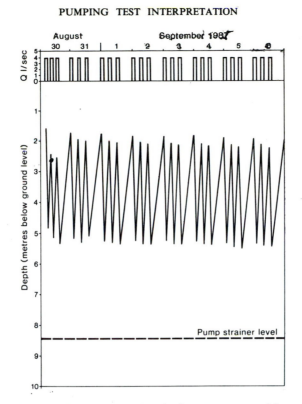

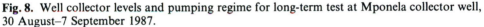

Fig. 8. Well collector levels and pumping regime for long-term test at Mponela collector well, 30 August–7 September 1987.

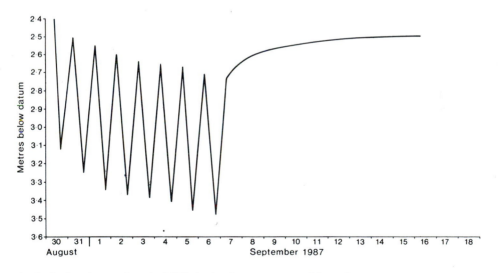

Fig. 9. Daily observations in MP1 during long-term test, Mponela.

4.2. St Nicholas School, Mahusekwa, Chiota, Marondera–Zimbabwe

Wright *et al.* (1986–7, Appendices Ia–IIb) give a full description of the dug well related work carried out at St Nicholas. As in the earlier example, the work down included geological surveys, site investigation holes and their testing, geophysical surveys, the construction of a dug well which was later converted to a collector well having four radials of a total length of 116 m and water chemical testing. In addition there was a full pumping test programme carried out and it is some aspects of this which are now restudied in a little more detail.

At this site the fresh bedrock interface with the regolith and saprock was relatively smooth over hundreds of metres. Seven exploration boreholes indicated depths to bedrock between 7.5 to 14.0 metres.

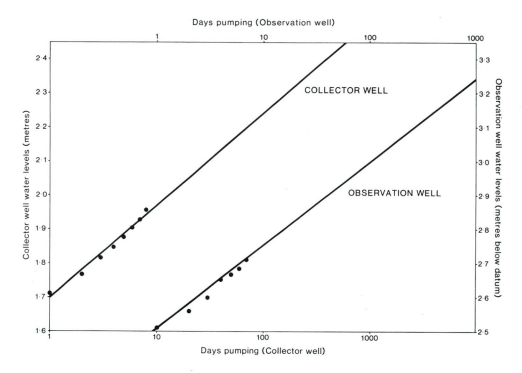

Fig. 10. Semi-log plot of water levels at pumping start, each day of long-term test.

The lithology was relatively simple at the dug well, which was installed to a total depth below ground level of 12 m. Only one material was encountered, reported to be sand with clay. It is possible that in terms of the idealized section of Fig. 1, the dug well fully penetrates the clayey saprolite confining layer and sits upon a very thin or absent band of saprock. Observations in the well showed water seeped gently into the well over the whole face of the clayey saprolite and that there may have been a relatively bigger inflow from the very bottom.

4.2.1.   The small drawdown dug wells pumping test, 19/02/86. The details of the field pumping test are given in Table 6 and the results determined assuming the confined aquifer condition are summarized in Table 7. Assuming $l = b = 9.48$ m and $K_H/K_V = 1$, the unconfined aquifer conditions for 50% recovery gives $T = 5.4$ m^2/d.

Table 6. *Details of the small drawdown recovery test at St Nicholas, 19/02/86*

Level of datum above ground level	0.65 m
Static water level below datum	3.168 m
Saturated aquifer thickness	7.89 m
Duration of pumping, t_p	180 min
Pumping rate	0.3 l/s
Well water level at t_p	3.776 m
Well drawdown, s_p	0.608 m
Time to 25% recovery	120 mins
Time to 50% recovery	308 mins
Time to 75% recovery	668 mins

Table 7. *Application of the confined aquifer nomograms to small drawdown test at St Nicholas School, 19/02/86*

	Transmissivity deduced (m²/d)		
% Recovery	25	50	75
	1.1	2.0	3.4
	Storage coefficient deduced (−)		
% Recovery	25	50	75
	>1	0.4	0.1

In using the analytical model of §3.3.3 for leaky aquifers, the following parameters were kept constant: $r_c = 1.0$ m, $r_w = 1.0$ m, $Q = 25.90$ m³/d and $t_p = 0.125$ days.

In the previous example of the use of the analytical model only four sets of time/drawdown data were used to obtain a best fit for the unknown parameters. In this case 24 sets of time drawdown data were used. These 24 data sets were read off the analogue hydrograph of recovery of well water level taken on site and were evenly spread in time from 0 to 1261 minutes after the start of the pumping.

In the first trial where best fit values for all of T(m²/d), $S(-)$, K_v(m/d), S_s(m⁻¹) and $S_y(-)$ were selected automatically, it was clear none of K_v, S_s or S_y had any effect on the result. It was therefore assumed the aquifer was single layered. A second trial, where K_v was accordingly set to 0, gave the following best fit results for T and S:

$T = 5.0$ m²/d, 50% confidence limits of 4.47 to 5.55.
$S = 0.062$, 50% confidence limits of 0.043 to 0.089.
(K_v, S_s and S_y do not affect the result).

4.2.2. The finite difference model used to simulate a large drawdown test carried out on the dug well. As for Mponela North, a finite difference model was constructed of the collector well aquifer system at St Nicholas. The modelling was done in two phases. The first phase was carried out before the results of the application of the analytical model were available. During this phase all the parameters were varied. As for Mponela, it was assumed during modelling that better hydrograph predictions were obtained when the model assumed a layered aquifer system. The specific yield was taken as 0.5 and the confined storage coefficient for the basal and confining layer was 0.001. With these selected values, various simulations were made of a deep drawdown test, some of which are shown on Fig. 11. Although the run, which assumed $T = 4 \, m^2/d$, gave the best agreement to the behaviour observed on site the recovery rate of the model is faster than that observed in the field.

The second phase of modelling consisted of one prediction only and assumed the aquifer was single layered having a T of $5 \, m^2/d$ and an S_y of 0.062. The results of the modelling are also shown on Fig. 11 and surprisingly the prediction is sufficiently close to that observed on site to suggest the aquifer could be single layered and unconfined, thus confirming the result of the analytical model.

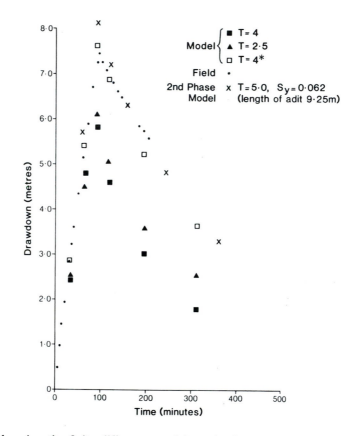

Fig. 11. Trials using the finite difference model to simulate a deep drawdown test on the collector well at St Nicholas, 17 July 1986.

4.2.3. Conclusions from the results of the short-term tests. Application of the (least-squares) analytical model shows that the best fit found was for a single layered aquifer having a transmissivity of $5 \, m^2/day$ and had an S_y of c. 0.06. This result was tested successfully using the finite difference model, also the Herbert–Kitching model for single layered unconfined aquifers gave a transmissivity of $5.4 \, m^2/day$. However, it should be noted that if the analytical model had not been available it would have been difficult to choose between the results of the two field techniques (Herbert–Kitching and Nomogram).

4.2.4. *The long-term test at St Nicholas.* As for Mponela North the dug well was eventually converted into a collector well. Four horizontal and four inclined adits were drilled to a total length of 204 m and a long-term test was carried out. This test was delayed until late in the dry season in order to obtain a better idea of how the well would perform with water levels at the seasonal low. The following data were recorded when carrying out the long-term pumping test:

Static water level: 5.403 m below datum
Pumping regime: Daily 06.00–08.00; 11.00–13.00 and 16.00–18.00
Pumping rate: 1.5 l/s

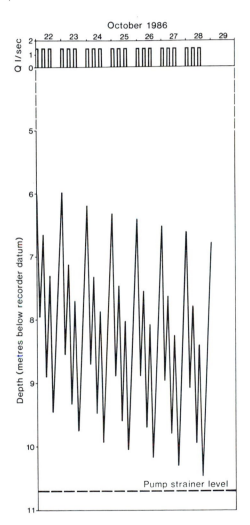

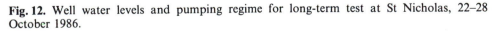

Fig. 12. Well water levels and pumping regime for long-term test at St Nicholas, 22–28 October 1986.

Well water levels recorded during the test are plotted on Fig. 12. As for Mponela North a plot of the drawdown observed at the start of pumping each day against, log of time since start of pumping test, was linear, see Fig. 13. If Jacob's method is applied to this plot, as for Mponela, the transmissivity indicated is about 5 m²/day.

Assuming the length of the dry season is 182 days, the predicted maximum pumping level for a rate of 1.5 l/s for six hours daily will be *c*. 12.5 m, Fig. 13. For the same pumping regime the safe yield will be *c*. 1.4 l/s, providing no nearby hydraulic boundaries exist.

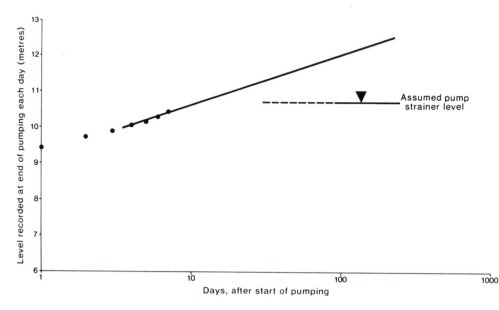

Fig. 13. Semi-log plot of water levels at end of pumping each day of long-term test.

5. Conclusions

New methods of pumping test interpretation specifically designed for dug wells constructed on hard rock aquifers have been demonstrated.

The new methods must be interpreted with care and more than one method may need to be applied to obtain a full interpretation of the hydrogeological system at any one location.

The weathered hard rock aquifer may be so variable that it is impossible to get anything but a very site specific result.

The following procedure is recommended for testing dug wells in such locations. Carry out the field test described in §3.2 and take a full hydrograph of the recovery of well water level to at least 50% recovery but ideally to 75%. The two interpretation techniques of §3.3.2 should then be applied. If these results are ambiguous the modelling techniques of §§3.3.3 and 3.3.1 must be applied.

The long-term test described in §3.4 will give a good indication of the maintainable yield during a long dry season. If this test cannot be carried out then the aquifer parameters found from the short-term test of §3.2 can be utilized in the usual way to make a prediction of long-term yield.

All the test interpretations described are applicable to aquifer systems other than weathered hard rock, providing the dug well is fully penetrating. A river alluvium, for example, should yield test results that are much easier to interpret.

Appendix: Notation

a	t_r/r_w^2
b	aquifer thickness
b_1	saturated thickness of a semiconfining layer
b'	t_p/r_w^2
c	$-0.127 \ln(b/r_w^T)$
d	$-0.332 \ln(l/b)$
F	$0.104 \, (b/r_w^T)^{1.02}$
K_V	vertical hydraulic conductivity
K_H	horizontal hydraulic conductivity
Q	rate of pumping
Q_0	pumping rate during a safe-yield test (§3.4)
Q_s	safe pumping rate
r_c	(internal) radius of the dug-well casing
r_w	(external) radius of the open (screened) interval of the well
r_w^T	$b(K_V - K_H)^{1/2}$
s_A	allowable drawdown
s_p	drawdown in the well at the end of pumping
s_0	drawdown at the start of daily pumping during a safe-yield test
$s_{0100}(s_{0180})$	drawdown (determined by graphical extrapolation) at the start of pumping on day 100 (180) of a safe-yield test
s_{0E}	increase in drawdown between the start and end of all pumping during a single day of a safe-yield test
S	storage coefficient of the aquifer
S_s	specific storage of a semiconfining layer
S_y	specific yield
T	transmissivity of the aquifer
T_1	transmissivity used in the finite-difference modelling case histories
t_p	period of pumping
t_r	recovery period (measured from the end of pumping)
a	Sr_w^2/r_c^2
λ	ratio of the volume of water taken from well storage to the total abstraction in period t_p (see Equation (1))
τ	$4\pi T s_p/Q$
ρ	t_r/t_p

References

BARKER, J. A. 1989. *Programs to Simulate and Analyse Pumping Test in Large Diameter Wells.* British Geological Survey Internal Report WD/89/23.

—— & HERBERT, R. 1989. Nomograms for the analysis of recovery tests on large diameter wells. *Quarterly Journal of Engineering Geology*, **22**, 151–158.

BOULTON, N. S. 1954. The drawdown of the water table under non-steady conditions near a pumped well in an unconfined formation. *Proceedings of the Institute of Civil Engineers (London)*, **3**, Part III, 564–579.

HANTUSH, M. S. 1956. Analysis of data from pumping tests in leaky aquifers. *Transactions of the American Geophysical Union*, **37**, 702–714.

HERBERT, R. & KITCHING, R. 1981. Determination of aquifer parameters from large diameter dug well pumping tests. *Gound Water*, **19**, 593–599.

HOLMES, A. 1966. *Principles of Physical Geology*. Nelson, New York.

TODD, D. K. 1979. *Ground Water Hydrology*. Wiley, London.

WRIGHT, E. P., HERBERT, R., MURRAY, K. H., BALL, D., CARRUTHERS, R. M., MACFARLANE, M. J. & KITCHING, R. 1988. *Final Report of the Collector Well Project 1983–1988*. British Geological Survey Report SD/88/1.

From WRIGHT, E. P. & BURGESS, W. G. (eds), 1992, *Hydrogeology of Crystalline Basement Aquifers in Africa*
Geological Society Special Publication No 66, pp 243–257.

Rural water supplies: comparative case histories from Nigeria and Zimbabwe

John Houston

Water Management Consultants, 2/3 Wyle Cop, Shrewsbury, Shropshire SY1 1UT, UK

Abstract. Two rural water supply projects based on groundwater sources are described and compared. One in Nigeria aimed to provide yields of greater than 1 l/s using electric submersible pumps to villages and towns with populations from 2000 to 30 000; the other in Zimbabwe aimed to supply yields greater than 0.2 l/s using hand pumps to small communities and villages from 100 to 2200 people. The higher yields in Nigeria are examined in relation to geology and aquifer characteristics, aquifer recharge, siting methods and borehole design. It is found that the higher yields are partly due to the criterion used to define a 'successful' borehole and partly due to other factors. Geology and aquifer characteristics appear to be similar in both places but recharge is much higher in Kwara State, leading to higher water levels and higher potential yields. A greater use of geophysics in Kwara State leads to a higher success rate and greater drilled depths into the bedrock in Kwara State also contribute to higher yields.

The two projects described are located in Nigeria and Zimbabwe, both being countries where considerable areas of Basement Complex are exposed (see Fig. 1).

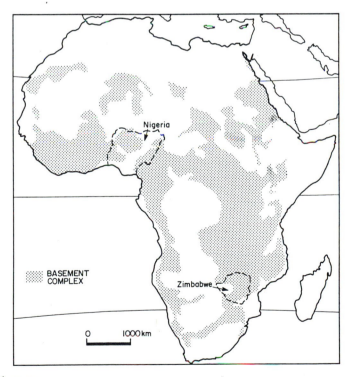

Fig. 1. Location map.

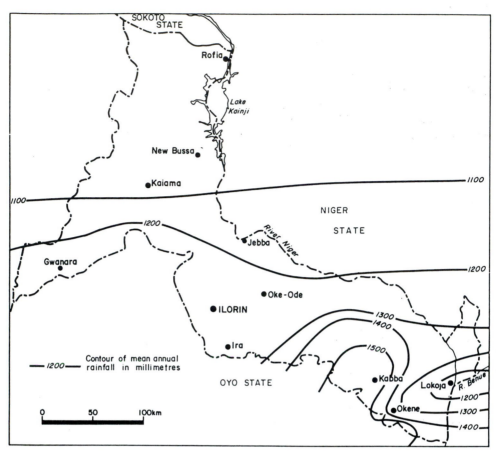

Fig. 2. Kwara State and mean annual rainfall.

The Kwara State Rural Water Supply Project in Nigeria (see Fig. 2) was carried out for the Kwara State Government by Biwater Ltd, with Hydrotechnica Ltd acting as drilling and hydrogeological consultants. The majority of water was supplied from groundwater and this part of the project took 15 months from 1985 to 1987. The demand centres were the major villages and towns spread throughout the State. The population of these sites varied from 2000 to 30 000. In all cases the supply was provided by electric submersible pump and was subject to primary treatment and subsequently fed into a distribution network.

The Victoria Province Drought Relief Project in Zimbabwe (see Fig. 3) was carried out for the Government of Zimbabwe and the EEC with Hydrotechnica Ltd acting as consultants. All water was supplied from groundwater and the project took 10 months during 1983 and 1984. The demand centres were small communities and villages scattered throughout Victoria Province which had been suffering from drought during the preceding two years. The number of people in these communities varied from 100 to 2200. The total population ultimately serviced was 250 000 in 358 communities, an average of 700 per site. In all cases the supply was delivered by handpump and no treatment or distribution was provided.

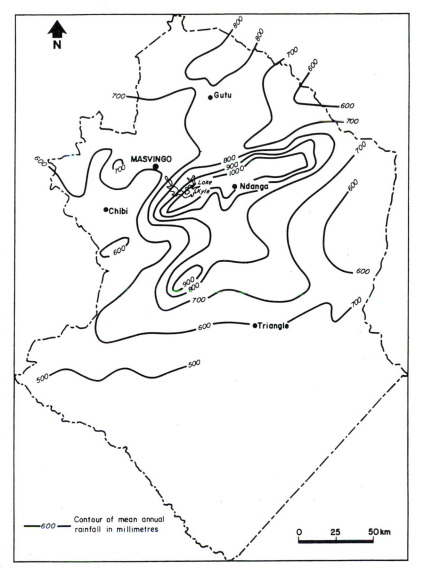

Fig. 3. Victoria Province and mean annual rainfall.

Thus the immediate aims of the two projects were somewhat different: in Kwara State the aim was to site a group of relatively high yielding boreholes in a wellfield environment, whereas in Victoria Province the aim was to site a large number of distributed boreholes with yields adequate for a handpump. Furthermore, the Kwara State Project was part of the longer-term planned development of the region, whereas the Victoria Province Project was a rapid response to a drought situation.

This leads to different criteria being used for the definition of a successful borehole and begs the question as to what criteria should be used to define a successful project. The definition of a successful borehole in Kwara State was that it should give a reliable yield of greater than 1 l/s, allowing for interference drawdowns between the boreholes in a wellfield, and allowing for a 12-hour pumping day during a 200-day dry season without recharge. This allows electric submersible pumps to be used and connected to a central power source with minimum pipeline lengths. In Victoria Province the definition of a successful borehole was that it should yield 0.2 l/s or greater to allow a hand pump to be connected which might be in operation up to 12 hours per day. This definition was modified during the programme to include the drawdown component and became a specific capacity of 0.003 l/s/m or greater, equivalent to a yield of 0.1 l/s with a 33 m drawdown. This reduction in specification led to a further 56 boreholes (15%) being equipped in areas of low yield potential.

Despite the disparity in immediate objectives between the two projects it will be shown that the results are remarkably similar when comparing the hydrogeology and that the main differences lie in project objectives and design.

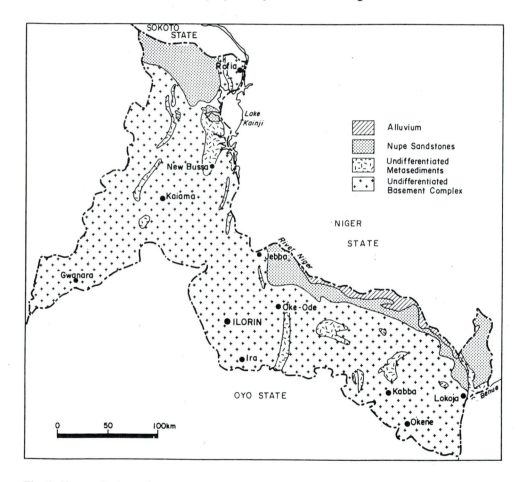

Fig. 4. Kwara State geology.

Geology and geomorphology

The geology of the two project areas is very similar: Basement Complex granites and gneisses (see Figs 4 & 5). The rock types in both Nigeria and Zimbabwe consist of granites, granodiorites, migmatites and gneisses. In both countries the older rock types are thought to be Archaean with later Proterozoic granite batholiths intruding. In Victoria Province, some of the gneisses (the Beitbridge Group) are part of the Proterozoic Mobile Belt.

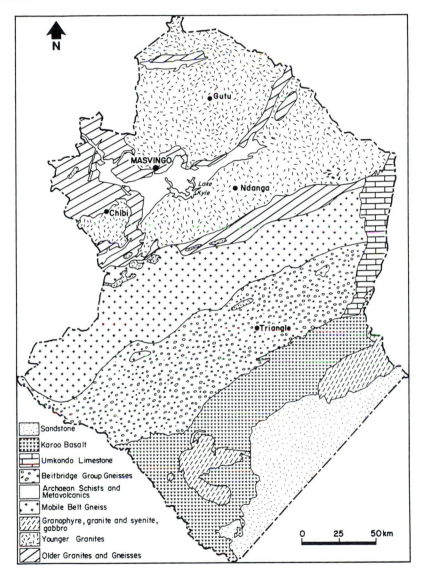

Fig. 5. Victoria Province geology.

No studies have been made to determine from a hydrogeological point of view whether there is any difference between the West African Craton and the Kalahari Craton but it is known that within Victoria Province the granites and the gneisses give rise to slightly different aquifer and hydrochemical properties (White *et al.* 1988; Houston & Lewis 1988),

Both Kwara State and Victoria Province are dominated by the polyphase Tertiary Post-African erosion surface (King 1962). A few remnants of the African Surface exist in Kwara State but are more extensively developed in Victoria Province. Conversely, in Kwara State there is an extensive development of the Quaternary flood plain of the River Niger, the only analogue in Victoria Province being the upper reaches of the River Limpopo flood plain which are limited in extent. Hence, the average elevation in Kwara State is around 400 masl compared with 800 masl in Victoria Province. The morphology of both areas is similar, with flat plains cut by shallow valleys and interspersed with bornhardts.

Aquifer characteristics

The aquifers developed in these environments are broadly similar and have been frequently described in detail elsewhere (e.g. Acworth 1987). They are composite, comprising a variable thickness of regolith overlying bedrock, the upper part of which is frequently fractured. Groundwater occurs throughout the high storage, low permeability regolith and the low storage, high permeability bedrock.

A comparison of the mean regolith thickness in Nigeria and Zimbabwe is given in Table 1. Perhaps surprisingly it is almost identical in both places. In Nigeria, however, the depth to water is consistently less, probably as a result of higher rainfall and recharge.

The physical properties of the aquifers in Nigeria and Zimbabwe have not been comparatively analysed but it is to be expected that permeability and storage are scale dependent, that is there is likely to be less difference between mean values in Nigeria and Zimbabwe than between adjacent sites in either area.

Table 1. *Comparative hydrogeological conditions*

	Kwara, Nigeria	Victoria, Zimbabwe	
	granite and gneiss[*]	granite[†]	gneiss[‡]
Mean depth to static water (m)	3.8 ± 2.7	5.1 ± 3.2	8.4 ± 6.6
Mean thickness of regolith (m)	18.1 ± 11.3	17.9 ± 12.1	19.2 ± 17.4
Mean thickness of weathered bedrock (m)			10.0 ± 7.2

[*] data from 165 boreholes
[†] data from 173 boreholes
[‡] data from 136 boreholes

Aquifer recharge

Rainfall varies over both areas (see Figs 2 & 3) but is generally greater in Kwara State. Where developed, the natural vegetation is heavier in Kwara State than in Victoria Province, leading to greater evapotranspiration in the former area. However, this effect is overruled on a local scale, since in the areas under development similar farming practices are in operation and little natural vegetation remains. This means that recharge estimates in both areas are based on comparable conditions. A variety of techniques have been used in each area to estimate recharge because none on its own is considered reliable. The techniques used are soil moisture balance models, baseflow analyses and chloride mass balance estimates. The various techniques give results which are in close agreement and strongly suggest that recharge in Kwara State is much higher than that in Victoria Province (see Table 2).

Table 2. *Comparative rainfall and recharge estimates for typical sites*

	Kwara,* Nigeria	Victoria,* Zimbabwe
Rainfall (mm)	1689 ± 436	736 ± 368
Recharge (mm)	253 ± 65	22 ± 25
Recharge/rainfall	15%	3%

* Data from 11 and 23 years respectively of soil moisture balance models.

Siting techniques

The method adopted for siting boreholes was much the same for both projects. An initial appraisal, taking into account whatever topographic, geological and hydrological information was available, would be made in concert with local social and logistical considerations. Once a general target area was identified a series of electromagnetic (EM) profiles would be measured to try and identify suitable geological structures for further assessment. The geological targets are generally thick zones of weathering or pronounced fracture systems. Since the resolution of such structures can be ambiguous with EM methods, specific sites would usually be further investigated using electrical resistivity equipment to obtain a vertical electrical sounding (VES).

In Victoria Province, detailed records of the methods used to site a borehole clearly demonstrate that success rate is increased by using geophysics to site boreholes (see Table 3). In particular where more than one geophysical technique was used in combination, very high success rates were achieved.

Whilst the siting methodology used was the same in both projects, it was possible to carry out more geophysics in Kwara State than in Victoria Province due to the time available (see Table 4).

Table 3. *Borehole siting success rates in Victoria Province, Zimbabwe*

Method	Successful boreholes (%)	Boreholes sited by method (%)
EM + VES	90	16.8
VES	85	34.7
EM	82	21.8
Hydrogeology	66	18.3
Air photos	61	7.5
Logistics	50	0.9

Table 4. *Comparative siting statistics*

	Kwara,* Nigeria	Victoria,† Zimbabwe
EM		
total line km	136.3	88.4
line km/borehole	0.64	0.24
VES		
total number	305	357
no./borehole	1.43	0.96

* data from 213 boreholes
† data from 370 boreholes

Borehole design

The approach to borehole design and construction was quite different between the two projects. Typical designs for both projects are shown in Fig. 6.

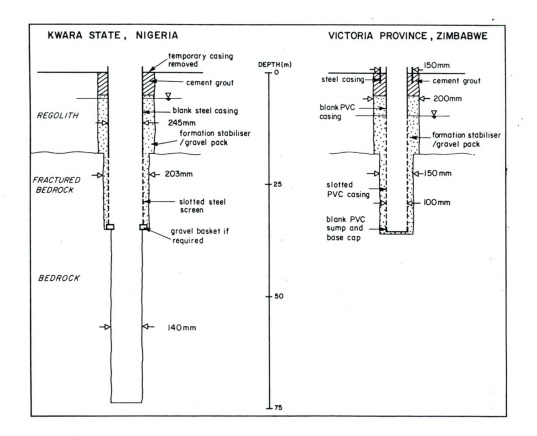

Fig. 6. Comparative borehole designs in Kwara State and Victoria Province.

Firstly, the requirement to be able to place electric submersible pumps in boreholes in Kwara State led to a decision never to complete at less than 150 mm diameter, whereas in Victoria Province the use of hand pumps meant that completed diameters of 100 mm were possible. The use of 100 mm as a nominal completed internal diameter has its own drawbacks such as difficulties in fishing operations, cleaning out and drilling on through to deeper depths, as well as prohibiting the use of higher yielding pumps in those boreholes which might otherwise supply more than hand-pumps. However, because yield is related to $\log(l/r_w^2)$ the larger diameter used in Kwara State boreholes will only lead to marginal increases in yield.

Secondly, the use of narrow diameter exploratory (test) boreholes in Kwara State allowed a preliminary assessment of a site to be made before committing to the greater expense associated with reaming and completing at production borehole diameter. Such an approach leads to greater flexibility in the programme but is only acceptable when the initial exploratory borehole is significantly cheaper than a production borehole. It is not realistic to use this approach when siting relatively cheap narrow diameter boreholes for handpumps. Whilst this method may lead to increased yields in production boreholes it will not improve siting success.

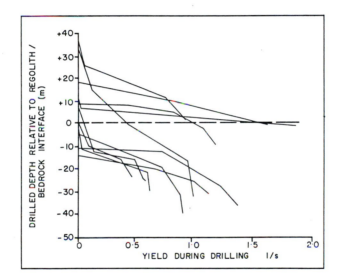

Fig. 7. Yield increases with depth during drilling.

Finally, yield increases with depth in the regolith tends to be fairly uniform, but, in the bedrock, yield increase is associated with specific productive fissures (see Fig. 7). The decline in the frequency of fissures with depth was demonstrated in Victoria Province (see Fig. 8). In Zimbabwe, 55% of the identified fissures were located within 10 m of the top of the bedrock and 87% within 20 m. Therefore, drilling beyond 20 m into bedrock should generally not be cost effective. Borehole depths were generally much greater in Kwara State (average depth 72.8 m) than in Victoria Province (average depth 35.8 m). This was the result of a deliberate policy to look for production from fissures at greater depths. In Nigeria it was recognized that in most cases the fissures and the majority of the yield came from the top 10 m of the bedrock. However, in 30% of boreholes the main yield came from fissures at depths greater (often considerably) than 10 m.

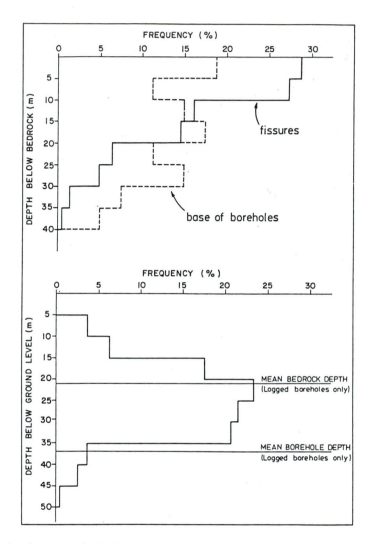

Fig. 8. Fissure frequency in bedrock.

Yields obtained

The yield frequency relationship for all boreholes in both projects (including those not equipped) is shown in Fig. 9. The relationship is similar, but when the results are aggregated (see Table 5) it is clear that in the lower yield range (0–1 l/s) there are 10% more boreholes in Zimbabwe and conversely in the higher yield range (1–5 l/s) there are 10% more in Nigeria. About 10% of boreholes in both projects gave yields above 5 l/s, suggesting that these occur randomly in any environment.

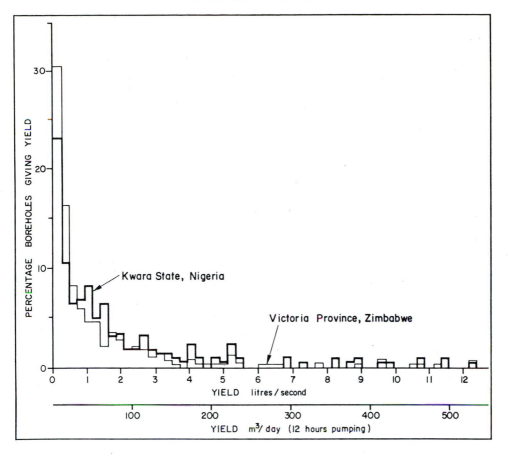

Fig. 9. Yield frequency in Kwara State and Victoria Province.

Table 5. *Comparative yield summary (percentage of boreholes giving yields)*

	Kwara,* Nigeria	Victoria,[†] Zimbabwe
0–1 l/s	55.1	65.6
1–5 l/s	35.6	24.4
>5 l/s	9.3	10.0

* all boreholes, including those not equipped: 213
† all boreholes, including those not equipped: 370

How comparable are these results? Unless the same method of measurement and recording is used it can be unrealistic to make any comparison. It is often the lack of uniformity in the raw data which gives rise to different results rather than actual differences in hydrogeology. In Nigeria, the yields given in this paper have been tested at or above the indicated rate for up to three days and the results extrapolated to a 12 hour–200 day safe yield, taking into account interference drawdowns within the wellfield. In Zimbabwe the yields were tested at up to a maximum of 1 l/s, because no other pumping equipment was available, generally for 3 to 4 hours. The decline in specific capacity with time (see Fig. 10) suggests that very little incremental decline takes place after the first hour and furthermore it is unlikely that a hand pump will be continuously in operation for more than a few hours. Therefore for yields up to 1 l/s the results are considered reliable. Yields above 1 l/s have been projected by multiplying the measured specific capacity by the available drawdown determined from drilling and geological logs (i.e. top of main producing zone minus static water level). Thus the data sets for each project are internally consistent, but differences still remain between the two projects and yield estimates for Nigeria may be slightly reduced compared with those for Zimbabwe.

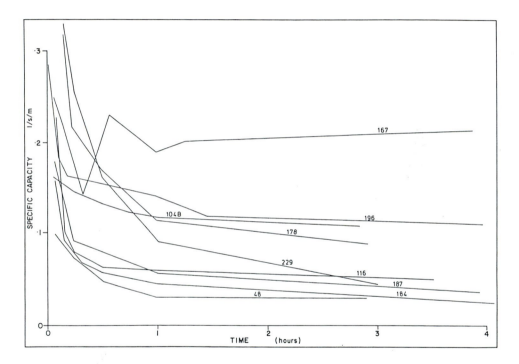

Fig. 10. Decline in specific capacity with time in Victoria Province Boreholes.

The probability distribution of specific capacity for all boreholes (including those not equipped) from both projects is given in Fig. 11. In the middle range, where the results are most comparable specific capacities are around 30% higher in Nigeria. In the lower range specific capacities are much higher in Nigeria. This may be due to the lack of accuracy in measuring both yields and drawdowns in boreholes which were going to be rejected. In the upper range, specific capacities in Zimbabwe are higher than in Nigeria and this may be due to errors in extrapolation noted above for high yielding boreholes in Zimbabwe.

A summary of the statistics for production boreholes for both projects is given in Table 6.

Table 6. *Comparative production statistics*

	Kwara, Nigeria	Victoria, Zimbabwe
total yield (l/s)	435.6*	359.8*
total drilled depth (m)	15504[†]	13248[†]
mean yield (l/s)		
arithmetic and (geometric)	2.64 (1.39)*	1.14 (0.56)*
mean depth (m)	72.8*	35.8*
mean drawdown	18.6*	12.1*
mean specific capacity (l/s/m)		
arithmetic and (geometric)	0.142 (0.074)*	0.094 (0.046)*
mean specific capacity per depth (l/s/m/m)		
arithmetic and (geometric)	0.0019 (0.0010)*	0.0026 (0.0013)*

* Production boreholes only (Nigeria = 165; Zimbabwe = 313).
† All boreholes, including those not equipped (Nigeria = 213; Zimbabwe = 370).

These results demonstrate clearly that production wells in Nigeria had more than double the yield of those in Zimbabwe.

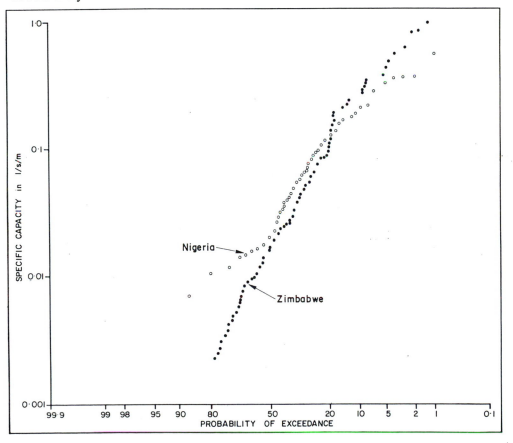

Fig. 11. Probability distribution of specific capacities in Kwara State and Victoria Province.

The results also show, however, that specific capacities of production wells were only about 50% higher in Nigeria and specific capacity per metre of well drilled was actually lower in Nigeria. The latter is clearly due to the reduction in occurrence of productive fissures with depth.

Conclusions and discussion

A number of conclusions can be drawn from this comparative study.

1. Production well yields in Nigeria are much higher than in Zimbabwe because the definition of a successful borehole was much higher. The mean yield and specific capacity of all boreholes in Nigeria and Zimbabwe was much closer.

2. The geology and aquifer characteristics in both areas are not significantly different. In particular the regolith depth is remarkably similar.

3. Rainfall and recharge is very much higher in Kwara State than in Victoria Province. One effect of this appears to be the higher water table in Nigeria compared with Zimbabwe and consequent greater available drawdown. This must inevitably contribute toward the higher yields in Nigeria.

4. The use of geophysics increases success rates (as demonstrated in Victoria Province) and increases production well yields. Approximately twice as much geophysics per borehole was carried out in Nigeria compared with Zimbabwe and this must be a contributing factor toward the higher yields there.

5. In Kwara State production boreholes penetrated an average 55 m into bedrock compared with an average 18 m in Victoria Province. This undoubtedly contributed to the increased yields in Nigeria. However, since the rate of yield increase with depth drops off rapidly below 20 m into bedrock, further consideration must be given to the cost effectiveness of drilling deeper.

The practical implications of these conclusions are that a significant amount of geophysics should be used as an aid to siting boreholes and that boreholes should generally be drilled into bedrock.

The cost effectiveness of geophysics can be demonstrated in Victoria Province, where the cost of the geophysical surveys amounted to around 10% (Z$960 in 1984) of the cost of the total borehole cost (Z$9600 in 1984). The actual success rate achieved in the project was 76% overall. If a second geophysical siting team had been employed the success rate would have increased to around 90% and the effective cost of a successful borehole would have reduced to Z$8250. In Nigeria, the cost of the geophysical surveys was less than 5% of the total borehole cost, thus making its use even more cost effective.

The marginal increase in yield as a result of drilling deeper in Nigeria compared with Zimbabwe was around 0.2 l/s. The marginal increase in cost in Zimbabwe for drilling the same additional depth was Z$1750. This represents an equivalent cost of Z$8750 per l/s compared with Z$17142 per l/s for a new borehole. Thus it is clearly cost effective to drill to greater depths. Exactly how deep requires further evaluation.

I would like to thank Biwater Ltd for allowing me to publish the data from Nigeria, and the Ministry of Energy and Water Resource Development for allowing me to publish the Zimbabwe data.

References

ACWORTH, R. I. 1987. The development of crystalline basement aquifers in a tropical environment. *Quarterly Journal of Engineering Geology*, **20**, 265–272.

HOUSTON, J. F. T. & LEWIS, R. T. 1988. The Victoria Province Drought Relief Project: II. Borehole Yield Relationships. *Ground Water*, **26**, 418–426.

KING, L. C. 1962. *The Morphology of the Earth*. Oliver & Boyd, Edinburgh.

WHITE, C. C., HOUSTON, J. F. T. & BARKER, R. D. 1988. The Victoria Province Drought Relief Project: I. Geophysical Siting of Boreholes. *Ground Water*, **26**, 309–316.

Index